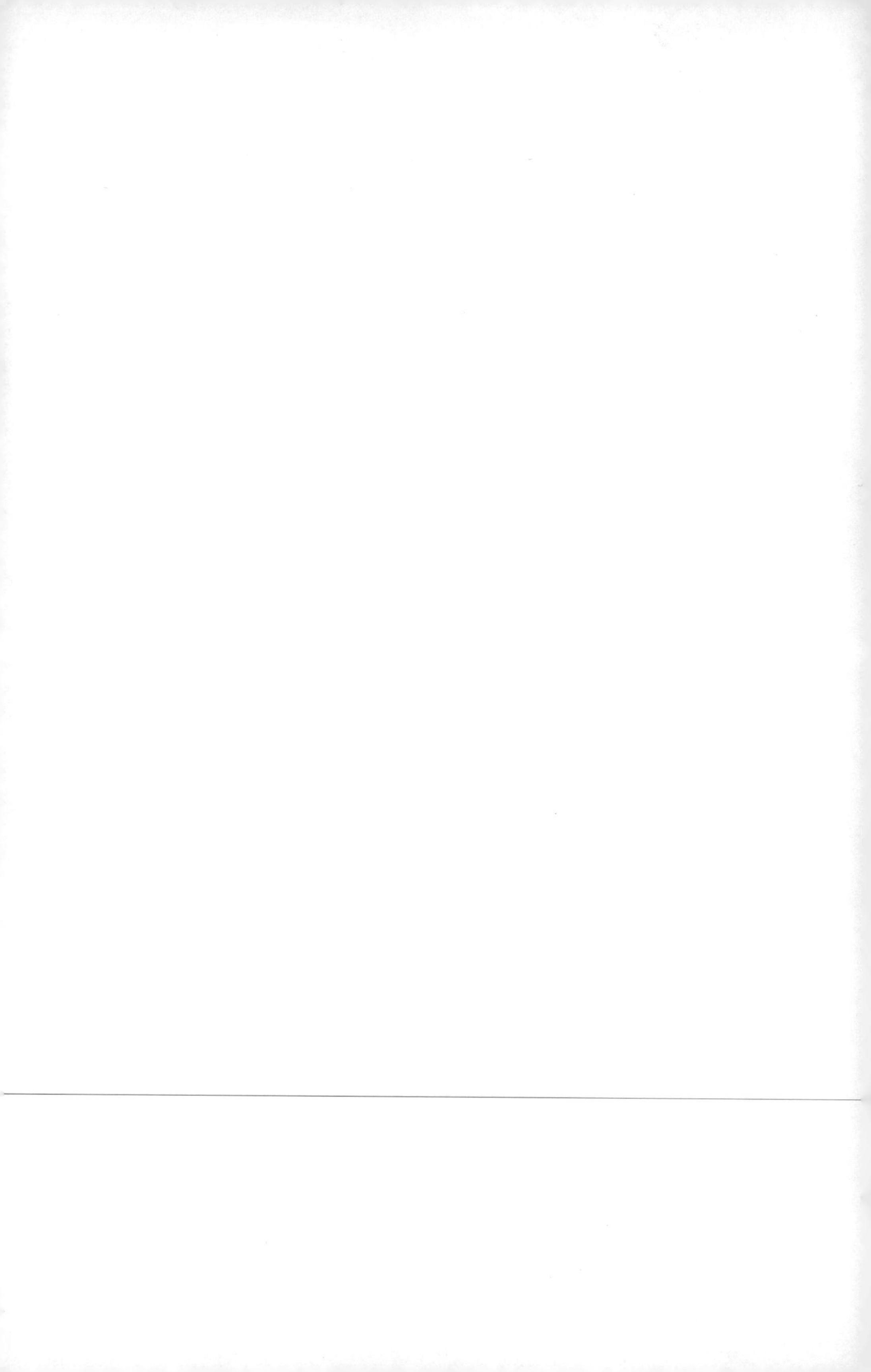

PRACTICAL ETHICS FOR A TECHNOLOGICAL WORLD

Paul A. Alcorn
DeVry Institute of Technology

Upper Saddle River, New Jersey
Columbus, Ohio

Cataloging-in-Publication Data
Alcorn, Paul A.
Practical ethics for a technological world / by Paul A. Alcorn.
p. cm.
Includes bibliographical references and index.
ISBN 0-13-660192-8
1. Ethics. 2. Technology—Moral and ethical aspects. 3. Applied ethics. I. Title.
BJ59.P73 2001
170—dc21

00-022959
CIP

Vice President and Publisher: Dave Garza
Editor in Chief: Stephen Helba
Assistant Vice President and Publisher: Charles E. Stewart, Jr.
Assistant Editor: Kate Linsner
Production Editor: Tricia Huhn
Design Coordinator: Robin Chukes
Text Designer: Ed Horcharik/Pagination
Cover art/photo: ImageBank
Cover Designer: John Jordan
Production Manager: Matthew Ottenweller
Electronic Text Management: Marilyn Wilson Phelps, Karen L. Bretz, Melanie N. Ortega
Marketing Manager: Barbara Rose

This book was set in New Baskerville by Prentice Hall and was printed and bound by R. R. Donnelley & Sons Company. The cover was printed by Phoenix Color Corp.

10 9 8 7 6 5 4 3 2 1
ISBN: 0-13-660192-8

To all my children, the ones of blood and the ones of life. To Jessica and Meagan, who have taught me compassion and the value of love given out. To Kimber and Tracea, who have taught me humility and the value of love received. To my students, who have given me the greatest gift, that of purpose. To all these I dedicate this book, with gratitude.

It is not on the Master that the "secret" depends but on the hearer. The Master can only be he who opens the door: it is for the disciple to be capable of seeing what is beyond.
Lama Yongden

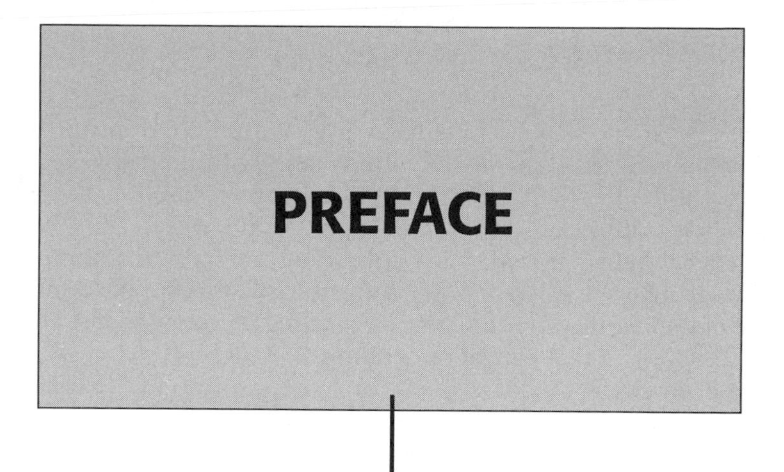

PREFACE

The purpose of this book is to create a guide—a body of information that can be applied to everyday life in order to enhance and simplify the decision-making process that everyone must go through in dealing with the ethical issues that are encountered daily. It is not a philosophy book per se, though certainly it discusses and presents various philosophies. It is not an intellectual treatise by design, though certainly it relies in the intellect as well as the intuitive powers of the reader to achieve its objective. This is a how-to book for people who have questions about how to live ethically, why a given course of action is or is not ethical, and how to operate in concert within their beliefs and the beliefs of society, thus experiencing success in this life—*in their own terms*.

This may sound like a great deal to expect from a single book, and a tall order to deliver for anyone. Yet it is my contention that the process of being and becoming ethical, much like the process of being and becoming human, is both natural and simply accomplished, if we only recall what we intuitively know about the nature of life. In order to accomplish this most difficult and simple of human developmental tasks, the text explores specific aspects of being ethical.

To begin with, we define the subject. In doing so, we briefly examine some of the common ideas people have about the nature of ethics, what the ethical life is like, and what being ethical actually means. We then take a commonsense approach to developing a somewhat different definition from those normally encountered and present it as the central theme of this work.

Next we look at the individual's point of view concerning the nature of the world. We concentrate on discovering the importance of recognizing one's own *paradigm* or world view and then work to discover how this influences the way we relate to our definition of ethics and the general theme of the book. Following this we study a few modern paradigms designed to explain human behavior in order to better understand our own motivations and examine some of the pitfalls of our paradigmatic structures—the way we see the world.

Then we examine how events in our lives are reflected in our behavior and what import that has on our ability and willingness to be ethical. Finally, to put the last piece in the theoretical puzzle, based on the information gathered to this point we will develop a general statement of how to achieve ethical behavior under any conditions and why this methodology works. We discuss throughout the book the ways all this applies to an understanding of the nature of technology, its place in our society, and how ethics fits into the fabric of that part of our culture.

I have been developing this book for more than twenty years. It is the culmination of a considerable amount of study and searching with the help of a great number of other people who all have the same question: *How am I supposed to behave in this life?* It would be much more than presumptuous of me to consider this my own work. I am little more than a reporter, recording the findings of those who have searched for life's answers for centuries, only to come full circle to find them within themselves. This is not a new idea or an old one that was developed by someone, found to work well, and then shared with others. This is a process of remembering what we already know on an intuitive basis. It is a process of cutting through the intellectual garbage to get at the truth, of separating the intellectual chaff from the grains of truth that lie within and setting them forth in a way that can be easily recognized and understood by today's people. It is, in fact, a process of shifting from a reality of rationalization—justifying our unethical behaviors—to a remembrance of the good sense with which we were born. Personally, I am delighted with the opportunity to share this material and the opportunity to create this volume, not because I have arrived at the perfection of ethical behavior (which I most certainly have *not*), but because we teach what we most need to learn. I can assure you that these are lessons that will increase your peace of mind, your happiness, your health, and your wealth more quickly and more completely than any other body of information at your disposal.

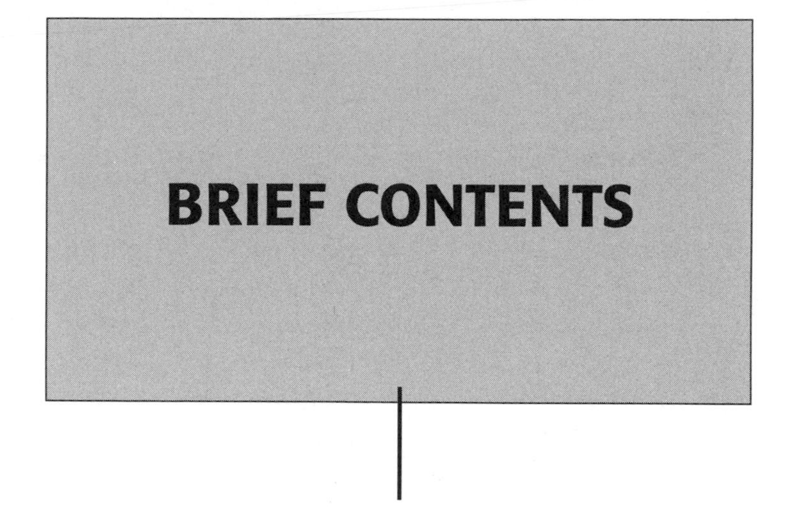

BRIEF CONTENTS

CONTENTS

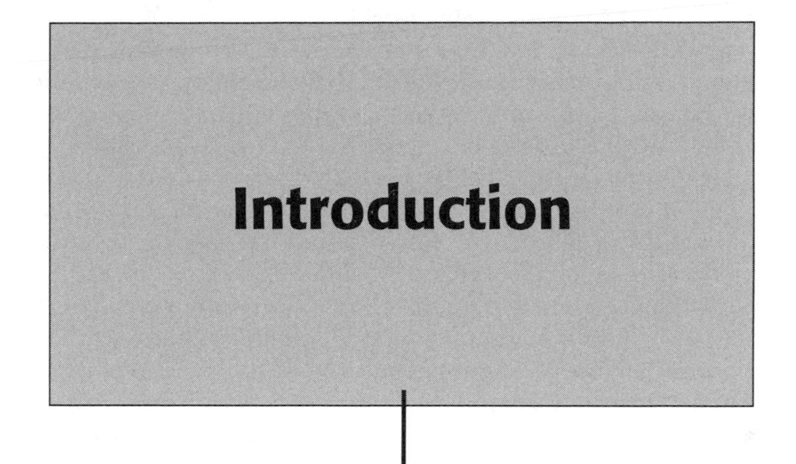

Introduction

The purpose of this book is to create a guide and a body of information that can be applied to everyday life to enhance and simplify the decision-making process that everyone must go through in dealing with the ethical issues that are encountered daily. It is not a philosophy book per se, though certainly it discusses and presents various philosophies. It is not an intellectual treatise by design, though certainly it relies in the intellect as well as the intuitive powers of the reader to achieve its objective. It is a how-to book for people who have questions about how to live ethically, why a given course of action is or is not ethical, and how to operate in concert with their beliefs and the beliefs of society and be ultimately successful in this life, *in their own terms*.

That may sound like a great deal to expect from a single book and a tall order to deliver for anyone. Yet it is the contention of this book that the process of being and becoming ethical, much like the process of being and becoming human, is both natural and simply accomplished, if we will only remember what we already intuitively know about the nature of life. In order to accomplish this most difficult and simple of human developmental tasks, specific aspects of being ethical will be explored.

To begin with, we will define the subject. In so doing, we will briefly examine some of the common ideas people have about the nature of ethics, what the ethical life looks like, and what being ethical really means. We will then take a commonsense approach to developing a somewhat different definition from those normally encountered and present it as the central theme of this work.

Next, we will take a look at the individual's point of view concerning the nature of the world. We will concentrate on discovering the importance of recognizing one's own *paradigm* or world view and then work to discover how this influences the way we relate to our definition of ethics and the general theme of the book. Following this, we will study a few modern paradigms designed to explain human behavior in order to better understand our own motivations and examine some of the pitfalls of our paradigmatic structures, that is, the way we see the world.

Then, we will take a look at how the past events of our lives are reflected in our behavior and what import that has on our ability and willingness to be ethical. Finally, to put the last piece in the theoretical puzzle, we will develop, based on the information gathered to this point, a general statement of how to achieve ethical behavior under any conditions and why this methodology works. Throughout, we will discuss how all this applies to an understanding of the nature of technology and its place in our society and how ethics fits into the fabric of that part of our culture.

This book has been developing for more than twenty years. It is the culmination of a considerable amount of study and searching on the part of not only the author but also a great number of other people, all with the same question: How am I supposed to behave in this life? It would be much more than presumptuous of me to consider this my own work. I am little more than a reporter, recording the findings of those who have searched for life's answers for centuries only to come full circle to find them within themselves. This is, in fact, not like some new idea, or old one, that was developed by someone, found to work well, and then shared with others. This is a process of remembering what we all already know on an intuitive basis. It is a process of cutting through the intellectual garbage to get at the truth, of separating the intellectual chaff from the grains of truth that lie within and setting them forth in a way that can be easily recognized and understood by the people of today. It is, in fact, a process of shifting from a reality of rationalization to justify our unethical behaviors to a remembrance of the good sense with which we were born. Personally, I am delighted with the opportunity to share this material and the opportunity to create this volume not because I have arrived at the perfection of ethical behavior (which I most certainly have *not*) but because we teach what we most need to learn. And I can assure you that these are lessons that will increase your peace of mind, your happiness, your health, and, as it happens, your wealth more quickly and more completely than any other body of information at your disposal.

Many thanks to Sheri J. Ironwood (DeVry Institute of Technology) and Ameeta Jadav (The Art Institute of Atlanta) for their helpful reviews.

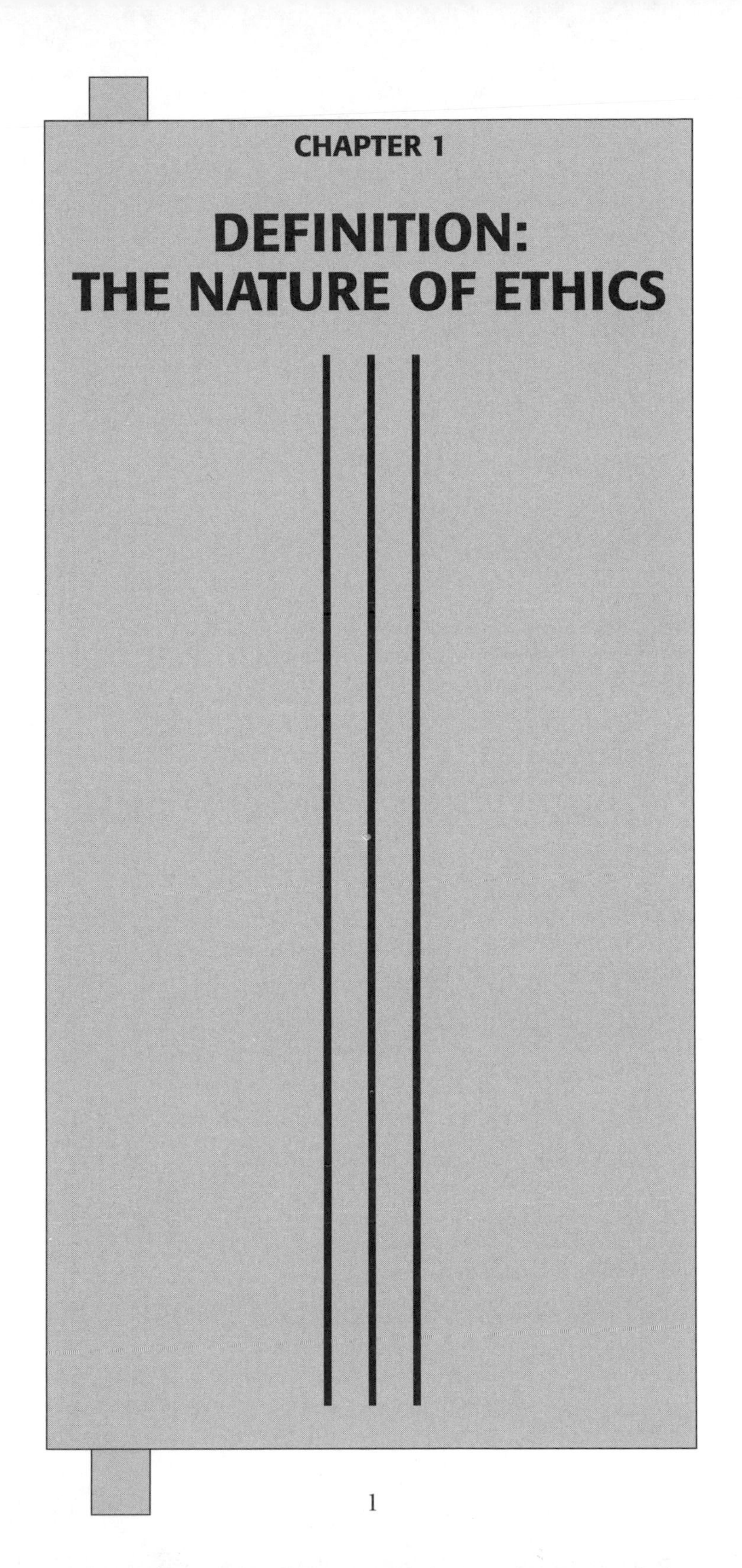

CHAPTER 1

DEFINITION: THE NATURE OF ETHICS

Thought in the mind hath made us. What we are
By thought was wrought and built. If a man's mind
Hath evil thoughts, pain comes on him as comes
The Wheel the ox behind . . .
. . . If one endure
In purity of thought, joy follows him
As his own shadow—sure.

James Allen

INTRODUCTION

So what in the world is ethics all about anyway? And why should I study it? Why should I be concerned? Isn't this just another way for somebody to tell me what I can think and what I can do? Who decides what's ethical and what is not?

A good place to begin is with what you *think* ethics is. I've asked this question of students numerous times, and in general, the answers fall into four specific categories: the dictionary definitions, the religious definitions, the "what I learned from Mom and Dad" definitions, and the experiential definitions. Surprisingly, most of these categories define ethics using similar terms and concepts.

Jot down your own definition of ethics, the way you see it now, to find out where you are at the beginning of this quest for understanding. Then, compare your definition with some of the statements that others have made concerning their understanding of ethics.

ALTERNATIVE DEFINITIONS OF ETHICS

1. *"Ethics is knowing the difference between right and wrong."* This is a pretty common response to the question. With the exception of sociopaths, virtually everyone develops in his or her own life some concept of right and wrong, and this is certainly not surprising, since everything from religion to the law deals with it. But is this a useful definition? Think about it. Suppose we ask, What is right and what is wrong? To answer that question, you have to either come up with further definitions to explain the initial ethics definition or list a number of examples of what is right and what is wrong and then determine how you decided their "rightness" and "wrongness." Probably, you'd find that it came from your childhood training, from interactions with parents, teachers, your big brother or sister, religious teachers, or other authority figures. In other words, it is an understanding of what is right and wrong based on what you have been taught as society's concepts.

This is what an anthropologist might call reproduction of the culture. Every culture will, by definition, reproduce itself in each new generation. Indeed, it is the conflict that exists in a culture between this tendency to reproduce itself and the desire and need of a culture to transform itself in the face of changing societal and ecological conditions that creates a balanced progress for that society. So in reproducing itself, the culture passes on information and concepts that have been found to be useful in the past. Hence parents teach their children the "rules of the game" for the society in which they live and mold their offspring into "moral" citizens who do what is right. The process is continued when you enter school; you learn what is and is not acceptable behavior in groups, what etiquette is insisted upon in social situations, which actions incur the wrath of teachers or fellow students and which do not, and so on. Again, it is a learned body of knowledge by which individuals discover what is expected of them and which actions will result in pleasure and which in pain. The process continues as well in the general society; certain modes of behavior are deemed unacceptable and others acceptable, each having recognized and mutually agreed upon consequences, ranging from praise for proper (right) behavior and ostracism for improper (wrong) behavior. For a society to function, it is essential that there be some mutually agreed upon sense of order. Our quest to maintain this stability and order is often tied to our concepts of right and wrong.

Yet are these learned concepts of right and wrong always ethical and appropriate, or are they culturally specific? That is, are they truly ethical in the universal sense or are they merely reflections of an individual culture at a specific moment in time? For instance, among the Maori of New Zealand, the traditional culture has some concepts of right and wrong that we might find rather bizarre. It is, for example, considered unethical to steal yams from your neighbor's garden by the use of black magic to entice them to migrate during the night into your own garden. To us this sounds ridiculous. Yet to a traditional Maori, such behavior is truly unethical and might result in serious social penalties.

Another example would be the practice among traditional Aleuts to share wives with guests, particularly during the winter, when the population is scattered and separated. Such behavior in our culture would be frowned upon, even in these liberated times, yet it is considered no more than good manners to an Aleut.

It would appear that using the knowledge of the difference between right and wrong as a method of determining what is ethical has some flaws in it. How do we know whether it is truly a matter of being ethical in the larger sense or merely a reflection of an individual culture? Is ethics really just a matter of local custom, or is there a body of knowledge that always applies? And who decides what is ethical—the individual, the family unit, the society in general? It would appear that we must look elsewhere for a useful, practical, and complete answer to our question, if one exists.

2. *"Ethics is seeking to be good and avoiding being bad."* Well, in a sense, we find ourselves in the same place as before, since usually our concepts of what is good or bad come from the same authoritative sources as do our concepts of right and wrong. Yet there is a subtle difference between what is considered right or wrong and what is considered good or bad. In the first case, the emphasis is on what is acceptable and what is not acceptable to society; in the second case, we have the added dimension of *judgment* on the individual. If a person does something that is wrong, he or she has made an error. The person is viewed as having broken basic rules and will experience negative consequences as a result. In the case of doing something bad, not only has the person done something that he or she should not do, but the act and the individual are viewed (that is, judged) as being bad. The subtlety lies in whether the person and the action are seen separately or as part and parcel of the same thing. Good and bad are statements of judgment against the individual as well as the behavior. Right and wrong are statements of judgment about the nature of the behavior in the eyes of society. We see people who perform bad acts as bad people who are less valuable than ourselves and unworthy of our respect. The person is being judged. What kind of person does something that is bad? A bad person, of course. And what kind of person does something that is good? A good person. Suddenly the perpetrator is seen in the same light as the deed, and being unethical becomes as much a matter of who you are as it is what you do. For some, this is not viewed as a problem, particularly if they are doing the judging, but for the perpetrator of an act considered to be bad, the fact that his or her behavior was bad does not necessarily mean that the person is bad. In other words, if we use knowledge of good and bad as a foundation for determining what is ethical, we are expressing an *opinion* about the event rather than the value of the behavior. We are passing judgment, and unless we think of our judgment as flawless, we must admit to the possibility of being wrong in our opinions. In that case, if right and wrong are the same as good and bad, we must be bad people to have been wrong in our judgment. What a wonderful way to get caught up in your own judgment! Again, the definition has flaws.

3. *"Being ethical means following God's laws."* The efficacy of this approach varies widely both in interpretation and perceived validity, depending on the religious background of the individual. Indeed, several major religions do not even involve a Supreme Being in their philosophy. Because of this, a caveat is in order at that point: Please be aware that just as this book is not really a philosophy book, neither is it a religious treatise. It does not purport to either criticize or support any specific religious point of view. Yet it is hardly possible to ignore the subject of religion in a discussion of ethics. That being said, let's proceed.

For many people, their religious texts and beliefs are the ultimate expression of how to live and what is and is not proper. That is, after all, one of the aims of religion—to teach the word of God, whatever that may mean to the individual practitioner. Yet there seems to be some disagreement among the citizenry of this world as to exactly what the word of God really is. Each religion and sect within religions has its own ideas about what following God's laws means. Those religions that have no godhead at all more closely resemble philosophies of life rather than what might traditionally be considered "religion" in our culture. Among these are Buddhism and the Eastern religions of Confucianism and Taoism, which rely more on the teachings of individuals and groups of individuals who present concepts for living a successful life than they do a relationship with God. Does this mean that there is no value to the concept of following "God's laws"? The problem with the argument lies not in the concept of "God's laws" but in the variance in interpretation as to what God has designed. Actually, as will be explained later, there is little difference between what different religions say in terms of ethics. They may differ in matters of tradition, ceremony, explanations of God's nature, and interpretation of events, but as far as what is and is not ethical is concerned, they pretty much agree. So where's the problem? It appears to be with the practitioners of particular religions who see their approach to religion as the only true understanding of God and hold the others as false. If this is the case, then it is not enough to say being ethical is following God's laws. We must also define whose version of God's laws is the correct one, and there appears to be a very wide variation of opinion among the religious of this planet as to which version to go with. In that sense, the definition of ethics is neither complete enough nor basic enough to offer a succinct understanding of the concept.

4. *"Being ethical means doing what's best for me."* Actually, there is considerable value in this definition, though probably not in what most people who say this really mean. This is what I call the hedonistic approach to ethics, whereby one reduces what constitutes ethical behavior to a matter of personal gratification. Another way of saying it is "If it feels good, do it." As selfish as this may sound, a quick look at the behavior of individuals in general will probably reveal that their actions are designed to create happiness for themselves. Adam Smith speaks of this approach in *The Wealth of Nations* as a matter of self-interest, which he sees as a necessary and natural way of behaving. If we really analyze it, we may even find that he is right. Everyone acts out of self-interest. At a very basic level, all actions are designed to further our own self-interests. Even the most altruistic of acts is a matter of self-interest if we consider the motivations of the altruist. Why else would anyone do something for another person? Isn't it because we feel we'll be better off in the process? As an example of how contraintuitive this concept may be, consider the case of the San Francisco policeman who risked his life to prevent a suicidal person from jumping off the Golden Gate Bridge. In reaching out for the falling man, he ran a very great risk of being pulled over with him into the water below and killed. He was successful in saving the man's life. Now that hardly seems like a selfish act on the policeman's part. Yet when he was asked about the incident later, he said that he was both frightened and aware of the risk of death but that he couldn't imagine not reaching out for the other man. He would have had to live with the knowledge that he might have been able to save another human being but failed to try, a prospect he could not face. The act of risking his life, in this context, was indeed an act of self-interest. So we can see that even the most altruistic and selfless of acts are, in a sense, for our own well-being.

Is this what people mean when they say that their ethics are based on doing what's best for themselves? Not usually. Often it is an admission of determining what is ethical strictly on the basis of self-betterment, even to the exclusion of others. Such a concept is totally untenable and indefensible in the cultural sense, in the religious sense, and in the practical sense. From the point of view of psychology, it would be viewed as an indication of a psychotic personality, that is, one who is unable to make decisions in the light of consequences, acting without morals at all. As we shall see, it is also in opposition to the overall thesis of this book and will, therefore, not offer a practical guide in and of itself.

5. *"Ethics is the study of what is moral and immoral."* At this point, we get a little closer to pay dirt, but still there are flaws in the argument. They stem from two main sources, one being that what most people think of as ethics is really a collection of statements about morality, and, therefore, the original definition really says ethics is the study of how to behave in an ethical manner. The other objection comes from the meaning of the word *moral,* which goes right back to what society and other authority figures say is right and wrong or good and bad. The word itself comes from the middle English *moralis,* which means "of manners or customs" and has the same base as the word *mores,* which refers to the customs of a given culture. We find ourselves once more mired in the muck of value judgments and personal/societal opinion. So apparently, though it may be an excellent guide to behaving within the accepted behavioral constraints of our society, unless you feel constrained to "when in Rome, behave as the Romans do" (or when living among headhunters and cannibals . . .), morality is not a good measure of ethics either.

Where does this leave us? Confused, perhaps, or somewhat overwhelmed? That's what too much philosophy does to a person. You may have noticed that it's much easier to refute any argument than it is to prove it. I hope that has just been illustrated. So now, to really confuse things, I'm going to present the definition that this book uses as its premise and add insult to injury by inferring that in reality, it agrees with all the earlier definitions offered, *if interpreted properly*.

A WORKING DEFINITION OF ETHICS

"Ethics is doing what works." That's it. It is short, succinct, and very straightforward. In its present form, it is also as useless as the definitions preceding it. However, this definition bears explanation. It simply states that the only motivation anyone has for any action is that it benefits him or her, either directly or by virtue of "psychic rewards" gained by helping others. If this is so, and if behaving in an ethical manner is desirable, then it must also be something that is beneficial to the individual. It follows, therefore, that *being ethical works,* as it is to the benefit of the individual. What other reason would a person have for being ethical? Why should we bother? If it is not to our own benefit to behave in a given way, then behaving in that way will not work!

Quite often, we hear someone say, "I must do this" or "I've got to do that" or even "I ought to behave in this way." For them we could hypothesize that *"ethics is a matter of knowing and doing what you should do, ought to do, or have to do."* Variations on this response all center in the determination of what we *should* or *ought* to do and how to find out what that is and then do it. This approach speaks to a view of life bound in rules and regulations, a natural consequence of social structure, unalterably and indelibly immutable and set. Dismissing them as effective definitions of what is ethical is a matter of realizing one central fact: *There is nothing that you should do, there is nothing that you ought to do, and there is nothing that you must do.* In reality, there are only two things that you must do: die (at least that seems to be the case so far) and live until you die. These are inescapable requirements for human beings. But everything else is a matter of choice, and most people miss that. That is not to say that there are some things that would not be a good idea to do—because they would result in negative consequences—only that they are possible in a person's life. This gives us an amazing degree of freedom in our choices; much of what we hear as "oughtas, gottas, and can'ts" are in reality statements of other people's experiences and beliefs about what would be beneficial or harmful if they were done. The problem for the rational person is to determine what, of all the possible avenues of behavior open to them, does work for his or her benefit. This is not as easy as it may first sound.

How are we to know what works? How do we determine this? To find the answer, we must distinguish between the concepts of *ethics* and *morals,* as traditionally defined and as discussed earlier, as well as among these and another concept that we have not yet explored, that of *values*. As the principle is used here, your values are distinctly different from your ethics and morals in that they do not represent codes of conduct, mutually agreed upon modes of social behavior, or any other learned response to conditions. They are different in that they are innate. As the concept is used here, values are built in. You have them when you arrive on this planet, and they represent an intuitive sense of what is and is not going to work for your life. Unfortunately, rather than being encouraged, our understanding of values is usually overridden as we grow up, "traded in" for a set of social concepts about right or wrong that may or may not agree with our values. We learn to ignore the signals that our value system sends to us when these conflicts occur through the overriding instructions of our elders and peers. Intuitively we may know that some action is okay or not, that is, will work for us or not, but we are convinced through the authoritative pronouncements of others or the logical, rational arguments of others that we are incorrect in our feelings. And the older we get, the more we learn to not listen to our feelings and our values. This is a truly disastrous formula for anyone seeking to be ethical, because *it is in our values that our innate knowledge of what is right and wrong lies!*

In spite of the social training (brainwashing?) we receive, our values continue to feed us accurate information about how to behave. Unconsciously and, at times, consciously we strive to express our values in the world. We are always about the process of attempting to do what works best for us. Yet in the process of striving to be ethical, we seem to fail time and again, as is evidenced by the results of our behavior. It can be said of any behavior that if it didn't work to create peace of mind in our lives, it was, by the definition offered in this book, an unethical act. What's the cause? The direct answer is *error thinking.*

ATTITUDES, INTENTIONS, AND ACTIONS

Error thinking is a process by which we warp our responses to the world and thus suborn our original intentions through ineffective and destructive behavior patterns. Figure 1-1 demonstrates how error thinking affects the results of our intentions.

If our original intentions were being carried out, our values would tell us some act, such as affecting the world at *A,* is a workable mode of behavior, and we would simply act to do that. The result would be that what takes place in the world is indeed *A*. However, what separates our true nature, our values, from the behavior that we exhibit in the world is a collection of *attitudes.* We have developed these attitudes through training and learning, through what others tell us is acceptable behavior, along with all the learned experiences of our lives—all of our fears, beliefs, and observations of how the world operates. It all acts as a filter. Compare it conceptually to what happens when you put the end of a stick into a body of water. The water distorts the view of the world below the surface, and the stick seems to diverge from the intended path. Just as you cannot spear a fish in the water by aiming at its apparent location, neither can you hit a target of intention by aiming at its apparent location. In fact, in some cases, we find that to affect the world as we wish to, we must do the opposite. This is similar to the oriental concept that everything contains its opposite.

In the oriental approach to understanding the world, there are subtleties of behavior and understanding that do not readily present themselves in occidental thought. The oriental mindset tends to be more holistic than linear. That is, the tendency is to view the world in

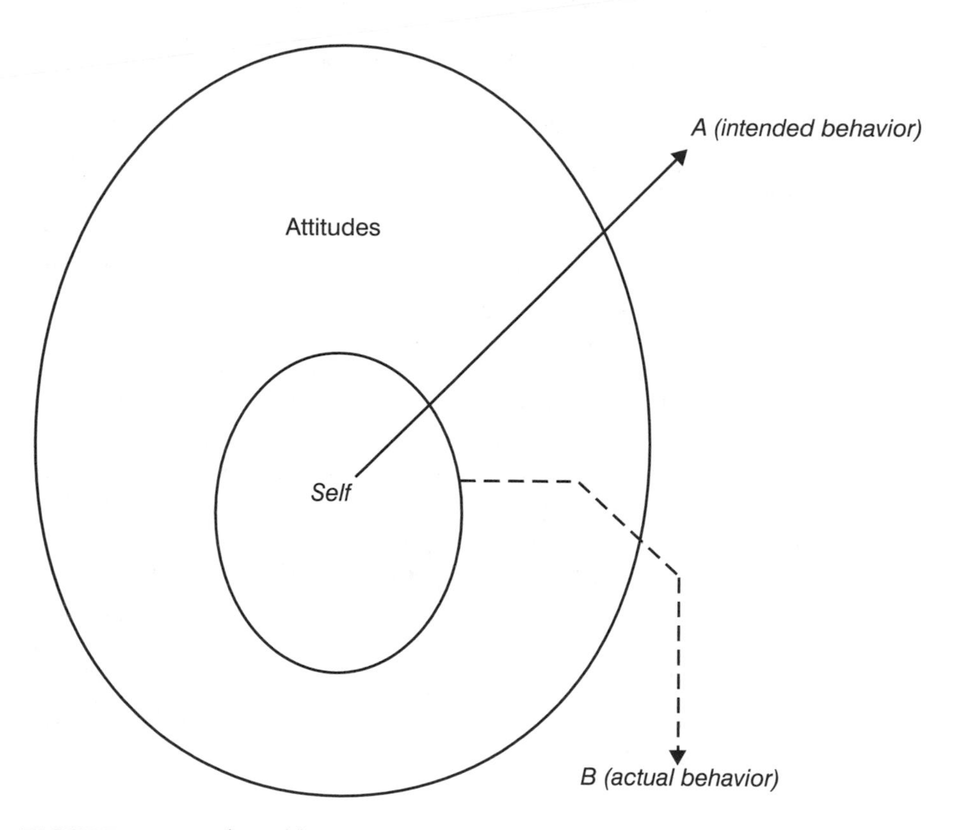

FIGURE 1-1 Real World

terms of a complete gestalt rather than a series of related cause-and-effect sequences and chains of relationship. As a result, there is a greater realization that everything has two balancing halves, each of which is necessary for the harmony of the whole. To consider action without taking into account the effect of that action on the holism of the structure results in trading one disharmony for another. Because of this, there are times when the most effective action is no action at all, or the most effective way to resist is to not resist at all. It is difficult to see this truth because of our learned attitudes. In truth, our actions are too often the result of attitudes rather than intention, and intention is not enough in and of itself. It must be coupled with a true awareness of the nature of the world, or what the Buddhists call *right thinking*. All of the attitudes that surround our true being distort our understanding of the nature of the world and the reality in which the target intention exists. The result is that we aim at accomplishing *A*, but through distortions we end up creating *B* instead. Even though we may have the *best of intentions*, it seems we miss the mark. The issue then becomes twofold: First we must determine what our values tell us is going to work and then deal with the distortions created by our attitudes so that we are capable of actually performing behavior that will work.

The key to being ethical is determining and knowing what works for your happiness, in your own terms. This is quite different from what most people imagine. Some people believe that having a lot of money would make them happy. Yet there are many people with money who are not happy and whose lives do not work. Another person might perceive happiness as a matter of being well thought of by others and think that this would bring him or her happiness. Yet there are those who are highly admired but who feel nothing but contempt for themselves. Others who are criticized and not liked by some might be perfectly happy with themselves.

Have you ever said, "I'd be happy if I could only . . . "? If you have and you received whatever it was you thought you needed, chances are you found it did far less to bring you happiness than you expected. Doing what works is difficult *because most of humanity does not understand what it is that works*. That is the true subject of this book. If it is possible to determine what to do to be ethical, you must determine what works for your own happiness, and that is at the same time totally different and exactly the same for each person. In the course of the following chapters, you will (1) discover what works for you, (2) begin to set up your life in a manner that allows you to achieve every goal you have, and (3) obtain every ounce of contentment that is possible.

COUNTERPOINT AND APPLICATION

On first inspection, the working definition of ethics presented in this chapter seems simplistic to the point of being useless. Yet it is actually a very functional definition to use in understanding the subject. In actuality, the definition is simple rather than simplistic, and there is an elegance in its simplicity that stems from the fact that it removes all of the verbose overkill usually attached to definitions for the purpose of making meaning perfectly clear. This definition has a slightly different purpose. It is simple in order to invite contemplation. It is free of qualifying verbiage in order to lead you to look within your own mind for interpretation to see whether you can make sense of it. It is on the order of a Zen koan—though not in the form of a question that causes illumination—not so much from the understanding at which you arrive as from the mental process of achieving that understanding. In truth, that simple statement, "Ethics is doing what works," speaks volumes, as we will see as we proceed.

Remember that the purpose of this book is not to present a historical or philosophical treatise on the ethical teachings of philosophers and logicians, though there is obviously great value in that process. Rather, it is to develop an applicable sense of ethics that you can carry with you for the rest of your life, regardless of what philosophy is in vogue or what explanation of the machinations of our physical lives are adhered to at any point in time. As for application of the principle, that is the meat of this entire text.

When we apply the principle to technology, the truth of the statement becomes self-evident. Technology that works, that is, does what it is designed to do, is by definition working. If it does not, it is not working. Accordingly, we could begin by saying that technology that works is ethical, as far as the device in question is concerned. That does not mean that its use is ethical, that the way it was created is ethical, or, for that matter, that the people who created it are useful. All those issues remain to be explored. As far as the technology itself is concerned, it is designed to do a job. That is the first element in determining the ethical nature of a technology. As we will see, this is certainly not enough to determine whether a technology is truly ethical; we should ask what the technology does in the process of performing as designed. Just because it fulfills its original intention does not mean that there are no unexpected or unintended consequences connected with its use. By extension, that also means that our definition of what works may need to be expanded, as we will see in subsequent chapters.

EXERCISES

1. The discussion of the five categories of answers to the question of ethics was not designed to indicate that the content of a person's moral training is incorrect but merely that the reliance on stock answers for the determination of ethical behavior doesn't work. The actual advice is often quite good. To illustrate this, revisit each of the five answers and note how you personally have depended on each approach at some time in your life and probably continue to do to some degree.
2. Take each approach in turn and think of ways in which you apply it. Then note how the application of determining ethics is manifest. That is, for each definition of ethics, how does that definition affect behavior and problem solving?

3. Make a list of statements of right and wrong actions, statements of good and bad behavior, examples of God's laws regarding behavior, examples of what you do and don't do because it makes you happy, and statements of moral and immoral behavior. Now look at the lists and see how many of the items actually work in your life. Are there some that you follow simply because you were taught to? Are there elements here that actually work well and, thus, you would do them anyway? Check the items in each list with either a *W* for "Works" or an *L* for "Learned." Look back and see how much of your behavior consists of rules and regulations and how much consists of a sense of what is really ethical and works for you.
4. Think about instances in your life in which you have attempted to achieve some goal and were met with abject failure. Try to determine how your actions in striving to achieve that goal were self-defeating. What could you have done differently? How did you sabotage your own efforts? Did you have any fears that led you to carry out some action that didn't work?
5. Consider people whom you judge as behaving in immoral ways. How are their behavior and beliefs different from your own? How much of your judgment is a conscious, rational understanding that what they are doing is not working and how much is a matter of a belief system that says they are (makes them out to be) wrong because their belief systems do not agree with your own?
6. The part of your answer for exercise 5 that reflects a belief in someone else's behavior as inappropriate because it differs from your own view is called intolerance. How intolerant are you of opposing points of view? Could it be that different approaches work equally well for different people? To put it another way, is it okay with you if people think differently from the way you do, or do you consider it an indication of moral and ethical inferiority?

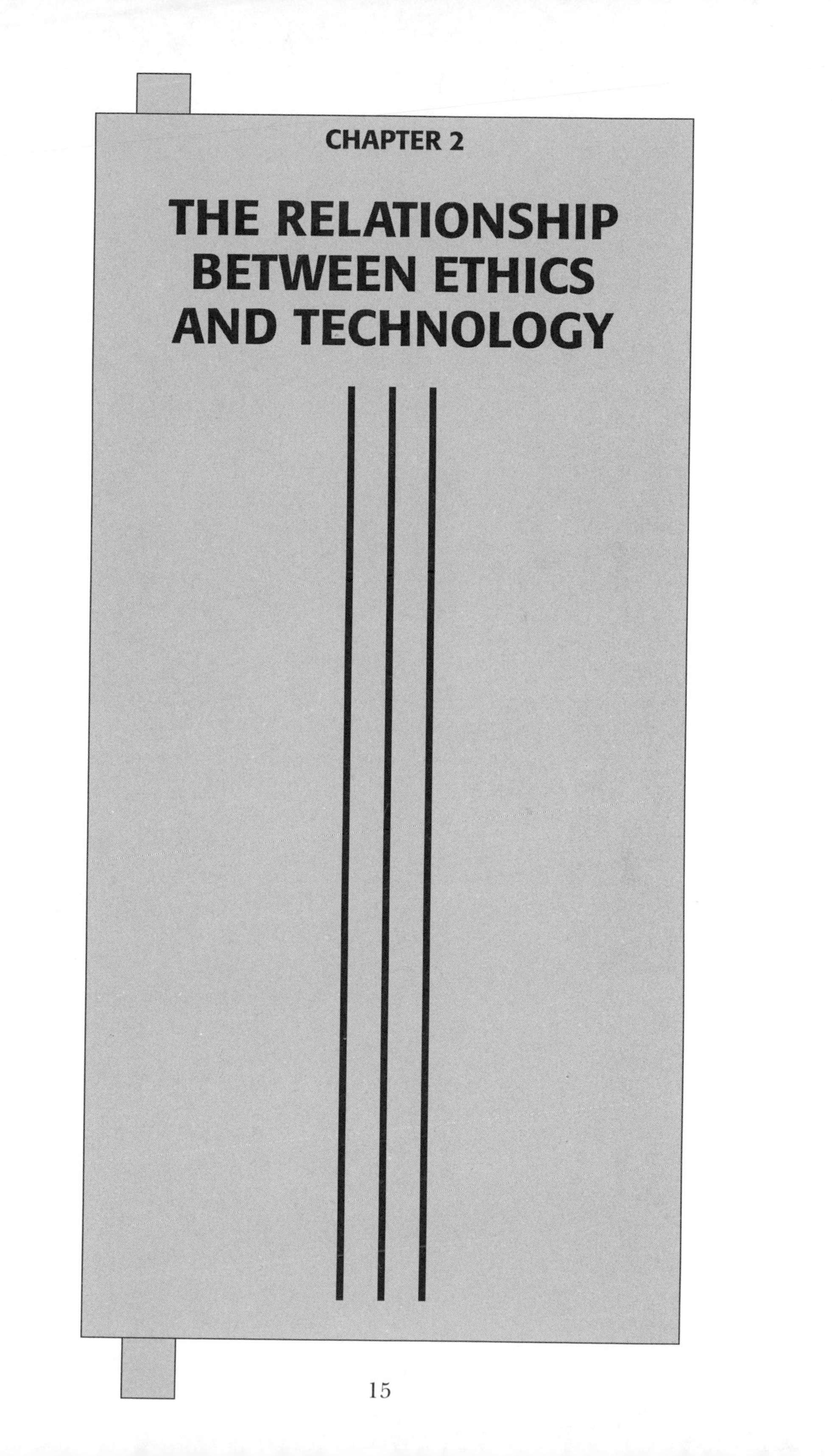

CHAPTER 2

THE RELATIONSHIP BETWEEN ETHICS AND TECHNOLOGY

I did not move a muscle when I first heard that the atom bomb had wiped out Hiroshima. On the contrary, I said to myself, Unless now the world adopts nonviolence, it will spell certain suicide for mankind.

Mohandas "Mahatma" Gandhi

In an evolving universe, who stands still moves backward.

R. Anton Wilson

INTRODUCTION

The first chapter offered a very short, general definition of ethics. In this chapter, we will develop a precise definition of technology to avoid misunderstanding when the term is used in this text. We will discuss the way in which technology is created and functions in a culture and then develop an understanding of how this concept of technology relates to ethical behavior. Finally, we will discuss how to use technology ethnically, that is, do technology in a way that works.

DEFINITION OF TECHNOLOGY

Essentially, technology is that whole collection of methodology and artificial constructs created by human beings to increase their probability of survival by increasing their control over the environment in which they operate. Technology includes and is essentially a means of manipulating natural laws to our benefit by constructing objects and methodology that increase our efficiency and reduce waste in our lives. The objects we create are artifacts, literally artificial constructs, that have been manufactured for specific uses and purposes. Everything that we use that is not as it comes to us in nature falls under the heading of technology. This is a very broad definition. All of the physical objects of our lives that were in any way altered from the way they appeared in nature represent technology. A sharpened stick is technology, as is a dollar bill or a caterpillar tractor; they merely have different functions and have been produced through a different series of steps, usually through the use of other technology.

It may be noted that human beings are not the only animals that create artifacts, and for that reason, the mere creation of artifacts does not in and of itself constitute technology. Birds build nests, chimpanzees use sticks as tools to gather food, and bees build elaborate hives. What is missing in these artifacts that separates them from what we mean by technology is the matter of choice. A bee contributes to the development of a hive because of genetic encoding. It is a process that is "hard wired," as an electrical engineer would say. It has no choice about what it is doing. The same is true of a bird building a nest or an otter using a rock to open a clam by resting the clam on its stomach as it floats and hammering it with a stone. Such behavior is instinctual. But not all methodology used by living creatures other than humans is instinctual. Some higher primates, chimpanzees, for example, are capable of reasoning through problems and using objects to create methodology for solving those problems. They have been observed experimentally under controlled conditions learning to attach telescoping rods together to gather food that is otherwise out of reach. Yet they have very limited capacity in this regard and do not pass this information on to others in a cultural way. What truly separates humans from the other members of the animal kingdom in this regard is our incredible power of choice.

TECHNOLOGY AND CHOICE

With humans, the technology we choose to build and the manner in which we use it is totally a matter of choice. We have an infinite capacity to produce technological goodies, within the boundaries of natural law, and we can accept or reject an idea as we choose. Thus, at one point in time, we may choose to develop the use of fire for cooking and at another decide to develop the art or science of architecture for the purpose of providing ourselves with shelter. Additionally, at one point we may decide to use dome-shaped hovels as shelter and at another time and place opt for alabaster palaces or multistory office buildings. The choice is all ours. It is in that choice of what artifacts to produce and the range of artifacts that we are capable of producing that we find the true nature of technology. And, as nearly as we can tell, that choice seems to be the sole province of human activity.

TECHNOLOGY AND EVOLUTION

In *Social Issues in Technology: A Format for Investigation,* I offered a detailed explanation of technology and the technological process. In this book I offer a general understanding of technology and why it exists in our lives. Technology is a vital part of what it is to be human; in order to understand our world, it is necessary to understand the purpose, the source, and the processes of our technological world.

For a human being, doing technology is a natural process. It represents one of the chief capacities with which nature has provided us for our survival. As with any other creature, Homo sapiens has certain characteristics that allow the species to perpetuate itself and successfully compete with other species for a niche in the natural world. Ecologically, we are an integral part of a much larger system that is designed to grow, develop, and maintain itself as an extensive living structure.

Every element in that system has the capacity to survive based on certain characteristics. For human beings, those *survival traits,* as these characteristics are called, include our capacity to create and use technology. There are specific and overwhelming advantages to this ability. Because we use artificial structures for our survival rather than develop the necessary characteristics through genetic alteration to our being, we are able to develop and adapt at a much higher rate than other animals or plants. We have effectively externalized the process of evolutionary development.

As an example, consider the characteristics of other animals versus those of a human being. Other animals have the advantage of speed, or claws, or special poisons that they can inject into their prey. Herbivores have specially designed digestive systems that allow them to consume large amounts of cellulose, a very difficult substance to break down, and turn it into useful energy. Some animals fly, others are very fleet of foot, others have incredible capacities to blend into the environment, and still others design complex living environments (e.g., hanging basket nests or colonized networks of tunnels). Each species has specific characteristics that offer it an advantage.

Now compare this with a human being. We do not have armored bodies covered with scales or shells. We cannot run particularly fast (though genetically we do have incredible stamina compared to most animals, a characteristic that allowed our hunter ancestors to follow game for days until the game was exhausted). Nor can we take to the air, with wings on our backs, or glide on membranes built into our bodies as bats or flying squirrels do. Yet we are capable of moving at a rate of speed far beyond that of a cheetah or other fleet-footed animal. We are able to fly across the face of the planet and into the outer reaches of our world and beyond. We can live underwater in craft that outperform the largest fish and exist in environments in which the extremes of temperature or altitude would kill most other creatures. We do it all in spite of the fact that we have at our disposal not a single physical trait that allows us to do so.

That is because the nature of our evolution has been external to our bodies. Instead of developing the eyes of a hawk, we develop binoculars and telescopes. Instead of becoming fleet of foot, we build automobiles and locomotives and airplanes. Instead of wings on our back, we have the wings of air transports and helicopters and the lifting power of balloons and dirigibles. Our characteristics are external to our physical being. It is in this ability to artificially create what we need for survival that we find our chief advantage. Like other animals, we use the laws of nature to aid us in our survival, but whereas other species do this through genetic alteration, a process that takes thousands if not millions of years, we manufacture the alterations quickly and efficiently. We find ourselves at last at a point at which we do not adapt to nature, we adapt nature to us! Such capacity is unparalleled in nature.

But with this capacity comes a problem. Nature is an experimenter. Nature will try numerous variations on a theme to find the combination of characteristics that allow a given organism to survive in a competitive world. If one alteration does not work, such as growing extra wings or limiting the number of eyes of a species to one, then that version fails and does not survive long enough to create progeny, or pass on the undesirable trait. If a variation offers superior opportunities for survival, many more of that version survive to pass on the characteristics to offspring, and eventually, that version predominates. Thus, through evolutionary mutation and survival of the fittest, we arrive at a creature that is perfectly adapted to its environment.

This is also true of humans, but with one exception. Since we are producing change through the creation of technology rather than trial-and-error mutation, we can very quickly generalize a new "trait" over the entire population in a relatively short period of time. In a matter of generations rather than millennia, a new technological device such as the bow and arrow or the chariot can come into general use by everyone who sees it. If it offers a very great advantage to those who have it, everyone either perishes or soon learns to use the new technology. There is little time for experimentation and testing here.

This has been seen often in the past with sometimes devastating results. The practice of agriculture is an excellent example if we look at the relationship between climatic change and the extensive use of agriculture in a region. Some of the most arid regions of the globe were once great forests or grasslands that were cleared for agriculture. Unfortunately, with the deforestation came a host of environmental changes that led to everything from soil erosion to changes in weather patterns. This is just a single example of the problems that can arise from moving too quickly to embrace a technology. Other examples include the virtual lack of forests in Lebanon today, where once stood vast woodlands of cedar, a prized wood traded all over the Mediterranean, from North Africa to Egypt to ancient Israel, and the cliff dwellers of the southwestern United States, who flourished toward the end of the first millennium and then abandoned their cities when they could not adjust to climactic changes in growing cycles.

What if the governments of the world in the last half of the twentieth century had decided that since nuclear weapons were the ultimate in destructive power, they would embrace that technology as is and abandon other means of war? We would have been left with no alternative but to create a nuclear holocaust in case of threat or attack. We are perhaps now in a similar predicament with biological and chemical weapons of mass destruction; they are cheap, effective, and easily produced and delivered. A single strain of a deadly bacterium or virus could cause a reduction of population around the world that would bring civilization as we know it to an end. And the tragic event would be the result of industrial and technological processes at work.

TECHNOLOGY AND RESISTANCE TO CHANGE

Because of this danger to our well-being, these seeds of destruction within our success, nature has also equipped us with another trait. That other trait is a resistance to changes in our culture. *Homeostasis*, as it is known, represents a fear of the unknown that extends to any technological device that may come along. Any new idea or new technology is initially suspect to most of the population because it is untested, unfamiliar, and therefore considered a potential threat. This is as much a survival mechanism as the capacity to create that technology in the first place. Because of homeostasis, time is a necessary ingredient for a given advance in technology to be generalized over the whole society. It is first embraced by a small section of the population eager to try new things and ideas, but the rest of society either initially ignores it or cautiously watches to see where it will lead. Should the new idea not be a particularly good one, that is, should it not increase the probability of individual and group survival, it tends to go by the wayside without much further ado. On the other hand, if it is actually a valuable idea, the new technology will continue to exist long enough for people to get used to it or to lose their initial fear of it, and then they are more willing to try this new gizmo. This is particularly true if those who first accept it have illustrated its value. Eventually, the acceptance and use of the new technology spreads throughout the culture.

This process can be easily seen in the case of the computer. Less than a century old, this device, once a curiosity used for certain esoteric operations by scientists and government, has become one of the primary tools of a modern technological society. It has been viewed as an oddity, feared, mystically couched in arcane terminology and given unrealistic assumptions of power by the uninitiated, seen as the subject of hobbyists and gadgeteers, embraced by big business, then small business, and finally accepted as an unavoidable way of life. The process took time while the population figured out how to use the new technology and how to configure it so that it was useful for their needs. It took time to gain acceptance and overcome the natural tendency of human beings to do things in the "same old way." It grew in popularity and use as a solution to a range of problems over the life of its development. All of that time was a gestation period for society to absorb and gain benefit from the new technology. Every invention goes through the same process, affected by a number of factors such as complexity, range of application, expense, and the degree of societal resistance.

The point to remember is that that resistance is necessary and natural, a safety net built into us by nature that allows us to take time to differentiate between new ideas that are truly beneficial and those that are potentially or truly dangerous to our survival. It is all part of the same natural process of creation and use of technology.

Human beings cannot help being creative. It is an element of our makeup that cannot be changed. Creativity and technological expertise require nurturance, but the tendency to learn the laws of nature and apply them to creating artificial constructs to enhance our lives comes as natural to us as breathing.

TECHNOLOGY AND ETHICS

Given that creating technology is natural and that within the limits of our understanding as to the nature of the universe, we can choose what technology to use and how to use it, where do the ethics of the process arise? If you remember back to our working definition, ethics is the process of doing what works. Apparently, from the history of the human race, using technology tends to work. This is evidenced, if in no other way, by the predominance and domination of our species over the face of the earth. We are incredibly successful as a species, reflecting incredibly successful natural traits, and that includes technology and its use. Apparently, technology works for us, or we would not include the capacity to create it in our repertoire of survival traits in the first place. By definition, then, in and of itself, it must be ethical.

That's a nice idea, and it would certainly be a blessing for all of us if that were true. Unfortunately, it is not as simple as that. Technology, as it turns out, is neither ethical nor unethical; it is merely a tool to be used or misused as we choose. Thus, we are back to the choice of action again, the one control we have in our lives.

Each technology and each application of technology raises ethical issues with which we must deal. Each new device or application of what we know requires some consideration of whether the use of that device will work for us or not. To further muddy the issue, we often cannot even say with certainty whether a technology will benefit us or not. In fact, in most cases, technology turns out to be a double-edged sword, with both costs and benefits in its use, and this in turn requires us to determine whether or not the benefits are worth the costs. And that's assuming we can even actually determine the costs accurately in the first place.

Also, we need to consider the idea that the use of technology may benefit some while costing others. This is not an uncommon occurrence, particularly where one technology replaces another, as in the case of the automobile replacing the horse-drawn buggy or the word processor replacing the "steno pool."

As you can see, this cost-benefit situation creates quite a dilemma. Just knowing that the ethical thing to do is to do what works is not very useful as a guide to behavior if we do not know what works in the first place. This is not a new idea. It is a problem that we as a species have been wrestling with off and on for ten thousand years or more, particularly when new technologies and new ways of manipulating the world present themselves. A few examples will clarify this point nicely.

When the automobile was first introduced, it was hailed not only as a solution to transportation problems within cities but also as a defense against growing pollution. That may seem quite confusing from our perspective as citizens of the world at the beginning of the twenty-first century, but a century ago, the pollution problems faced by industrial urban dwellers was decidedly different. At that time, at the birth of the automobile age, the chief means of transportation was the horse. Anyone who wasn't walking or traveling by train within an urban environment was traveling on foot or by horse. Carriages, drays (freight wagons), and specialized coaches were all horse drawn. With the horses came horse dung, and it was everywhere. The streets were pocked with piles of dung to be cleaned up, dung that ran into the sewers and that produced a prodigious number of flies. And with the flies came disease. We do not think of horses and horse dung as being a major health hazard in our lives today, but a hundred years ago, it was a major problem. Thus, the "horseless carriage" was hailed as the eliminator of the "hay burner" technology of equestrian transportation.

Yet today, we view the automobile as a chief air pollution source; it dumps tons of carbon monoxide and other pollutants into the atmosphere, promoting global warming and creating smog in any city of size. Hence the solution becomes the issue. At the present time, there is a movement toward nonpolluting electric cars. California has gone so far as to mandate a 10 percent noncombustion engine vehicle quota for the state. Electric cars are the obvious noncombustion engine choice, and as the number of electric vehicles rises, replacing gasoline engine automobiles, it is believed substantial improvement in the environment will result. And so another solution has been found.

This being the case, should we not expect these electric vehicles to create other dilemmas? At the present time, nearly all electric automobiles are powered by heavy lead-acid batteries, deep charged and able to deliver power at sufficient rates for a reasonable amount of time. And much research is being done to develop better and more powerful batteries that will charge more quickly and deliver more power for even longer periods of time. Thus it appears, at least for the foreseeable future, that a dependence on lead-acid batteries will be dominant. But a new problem arises: What will we do with the spent batteries? Batteries are already seen as a pollution problem, with only one per car. What will happen when the number of batteries per vehicle rises to twelve or twenty? Could we be exchanging one form of pollution for another? It is not just lead-acid batteries that present this type of dilemma as we progress and change technology.

With any technological change and any acceptance of a new technology as standard, there is always a cost. There is never a free lunch, though payment can be deferred for some length of time. Yet in the end, someone has to pay, and I'm sure it comes as no surprise that delaying payment until our children or grandchildren are making the rules is not a very efficient way to operate. Intuitively it is unethical to use this approach, though economically or politically it may be expedient.

To what extent should we consider the future payment for our exploitation of technology and technological possibility? Though we do not always know (indeed, seldom do we know) the true cost of a technological development, there are certainly some issues that we do know will need to be handled. History offers numerous examples of what to expect from technological change. How far does our responsibility go? One school of thought says not to worry about the future consequences because we have always been able to deal with what comes along. Still newer technology will solve the problem. New ideas and alternative ways of handling the issues will arise naturally out of necessity. We need only utilize what is available to us now, and let the future generations worry about how to handle the problems that arise. These are the attitudes that led to the destruction of environments in the ancient world. As agriculture and population exploded beginning some ten thousand years ago, whole civilizations were destroyed by resultant drought and crop failure. Whole ecosystems were altered, turning fertile plains into deserts and lush forests into arid wasteland. Solutions were found, but what was the cost? The people of these transitional periods endured starvation and being uprooted as their productivity collapsed.

On the other hand, consider the approach of the Five Nations of the Iroquois Confederation. These Native Americans of the northeastern United States banded together in a peaceful structure that allied independent nations, building a greater confederation. The Cuyahoga, Seneca, Onondaga, Mohican, and Oneida nations agreed to work together for the betterment of all and for their mutual defense against their unfriendly neighbors, chiefly the Algonquin. This amazing group of people elected fifty men from among their number to collectively make the decisions for the whole group. (Interestingly, it was the women of the tribe who actually chose the fifty men to head the joint council.) They always considered the future consequences of those decisions, *for seven generations hence!* No decision that was merely expedient was acceptable. Compare this approach with the political process present in most industrialized countries today. How many decisions are made on the basis of how the people will be affected a century and a half in the future? It appears we could learn a great deal from these Native American tribes. (Incidentally, it would not be a wise idea to embrace the wisdom of the Native Americans without exception. The Iroquois, for example, are noted for their horrific treatment of prisoners of war, whom they first honored and then tortured for as long as possible without killing them, then ritually ate them, not for the food value, but to absorb some of their bravery and strength. It was considered a pity if the prisoner could not be kept alive in a state of agony for at least twenty-four hours before he or she died.)

Numerous other examples can be cited describing the failure of humans to include negative future circumstances in their deliberations. Again and again we see in the industrialized world the adoption of a technology that results in future problems. This is not to paint a dark portrait of technology or to suggest that we should abandon our technological ways. Our whole history as a nation has been one of progress and growth. It is merely a reminder that every new opportunity brings with it an obligation to consider the consequences of our actions, and this we seem rather reluctant to do.

COUNTERPOINT AND APPLICATION

If technologizing, that is, creating and using technology, is so natural to being human, then it would appear that it is always an ethical process, as it always works. It is not very useful for any species to go against its nature in the quest for survival, except as a part of the evolutionary process, and natural selection would seem to be quite adequate to this end. Why all the fuss about the ethical nature of technology? It is neutral. It is what it is. Talking about the ethical nature of technology is like talking about the ethical nature of a stick. Isn't that true?

Of course, that is not true at all. It must be remembered that the drive to create technology and thus evolve externally to our bodies is indeed a natural process, yet it still entails free will, or choice, on our part. An almost infinite array of technological possibilities is available to us, depending on how we choose to apply the basic principles that constitute our understanding of the physical universe (physical laws). It is because of that choice that we must consider ethical content.

Surely, the homeostatic tendencies of the species goes a long way toward allowing us to adjust if we make mistakes in our choices in technological design and creation. But considering the speed at which the world changes and the far-reaching effects of even the seemingly most insignificant changes in methodology, it becomes critical to consider the usefulness of technological change in the broadest of terms, and that is a matter of what is the ethical thing to do. Technology has ethical content by virtue of the free will with which we create it. What do we choose to do and not do? We make those choices in a desire to improve our position in life, either individually, collectively, or both. Do we know that our choices are sound ones, and do they truly work to achieve the goals that they are designed to achieve? Therein lies the ethical issue.

When looking at technology and creating technological change through the modification, production, or application of technology, it is wise to think in a broader context. It is best to consider why exactly we are doing whatever it is we are doing, what our goals are, and whether the process undertaken actually achieves those goals. Additionally, we must consider what other goals or conditions are affected by the new creation or application and how that affects our overall goals in life. In other words, what would happen if we all behaved like the Native American confederacy mentioned earlier and considered the consequences of our actions for the next seven generations, or 150 years. How would we behave differently?

EXERCISES

1. Choose a technology with which you are familiar and that is generally used in your society on an ongoing basis. It could be anything from computers to airplanes to rubber gloves. Now begin to consider the positive reasons for the existence of this technology. Make a list of its benefits. Note how each is of value to the culture and why the technology exists as a method of doing what works.

   Next, consider the other side of the same technology. How does this technology impact you and others negatively? If you are having trouble finding anything negative about the technology you have chosen, think again. There is no such thing as a technology without a cost, and not just a monetary cost. How does the technology change your job and the jobs of others? How does it affect relationships and values? How can it be used for harmful purposes? Now compare the lists and determine whether the benefits outweigh the costs.

2. Consider ways in which the development of technology could be approached to create a more positive outcome for a society as a whole. If you were in charge, how would you change licensing and manufacturing regulations to better steer technology away from negative consequences? Is this a practical idea? Should we control the development and dissemination of technology because of possible negative effects? Is this concept in opposition to the concept of creative freedom, stifling potential new ideas for fear of censure? Considering that new ideas can

be controlled and kept secret through governmental and industrial forces, is this done to our benefit or our detriment?

3. It can be argued that every technology has both positive and negative possible consequences and that the technology itself is neutral. The truth may be that it is in the application that the ethical or unethical nature of a technology exists and nowhere else. To explore this, consider one of the following technologies and make a list of both the benefits and costs to us all as a result of that technology's use. Make the lists equally long, even if you have to cite potential but not realized uses or consequences of the technology.

a. Nano-technology

b. Urban development

c. Nuclear weapons

d. Dynamite

e. Helicopters

f. Mono-cropping agriculture

g. Printing presses

h. Computers

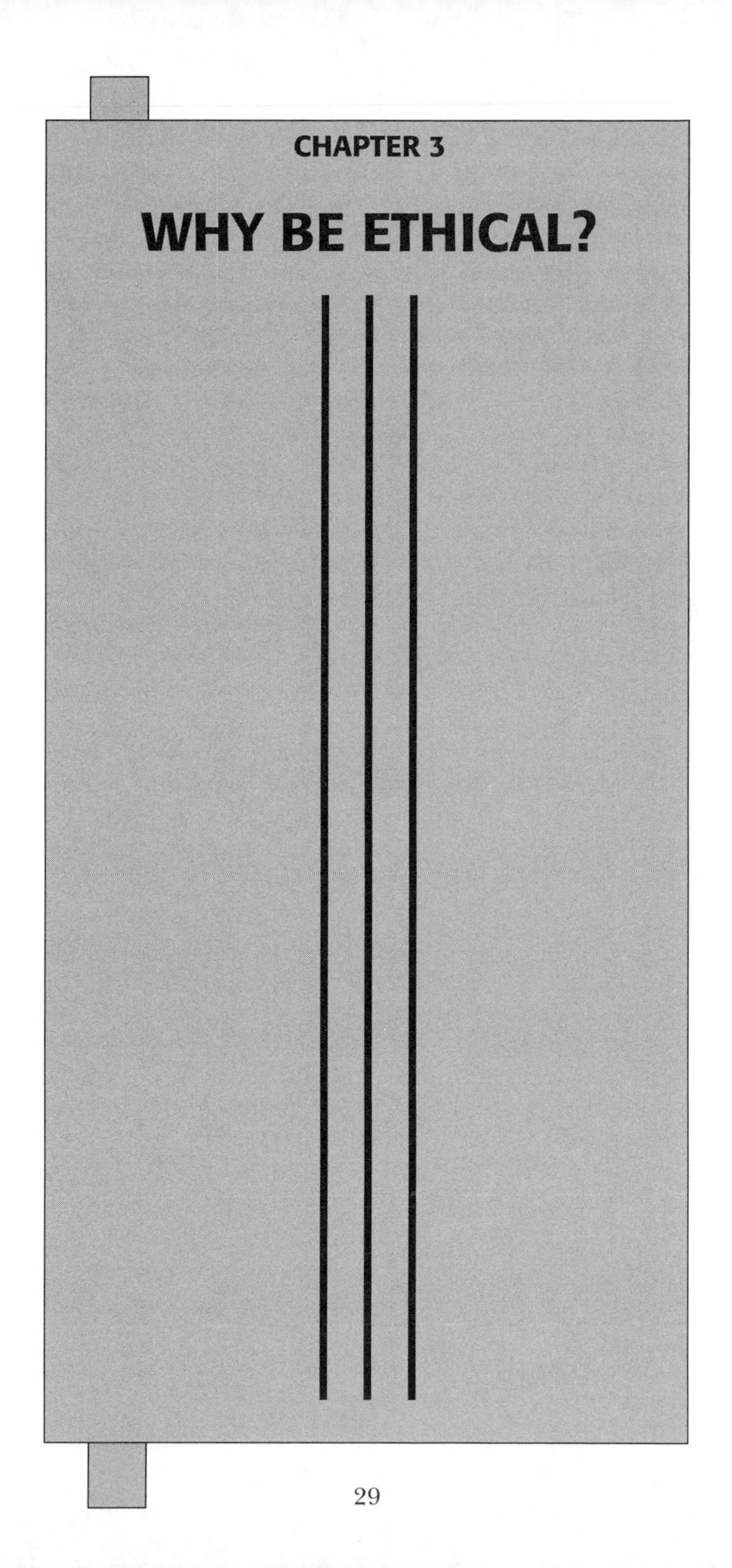

CHAPTER 3

WHY BE ETHICAL?

INTRODUCTION

You were introduced to the idea that being ethical was a matter of doing what works, in the sense that the individual, through ethical behavior, can achieve happiness on his or her own terms. Now all that sounds rather noble and exciting and, quite frankly, somewhat vague. What is meant by "happiness on his or her own terms"? What in the world is happiness anyway? In this chapter, we will explore some explanations for this expression.

If you ask ten different people for a definition of happiness, you will get a *minimum* of ten different answers. In terms of content, everyone has a unique idea as to what constitutes happiness. Yet, as the saying goes, everyone is the same only different. That is, everyone has a different way of finding happiness, though all are looking for the same thing. It is the contention of this presentation that when we talk about wanting to be happy, about finding true happiness, what we are really talking about is the quest for *peace of mind.* So for future reference, we will talk about peace of mind as the goal of our behavior. Logically, this is a fairly obvious truth. Even the self-interest discussed in the last chapter, the tendency to do only the things that we think will allow us to be better off, is an expression of seeking peace of mind. If we have no fears, we have peace of mind. If we have enough money, it gives us financial peace of mind. If we have the perfect job, with the perfect boss and the perfect working conditions, we have professional peace of mind. We are looking for just the right combination of conditions and opportunities to allow us to be content with our lives, and that will certainly lead to peace of mind. So how do we attain this goal?

You've probably heard people say that happiness is a state of mind, and in a very real sense, this is true. The ability to achieve and maintain peace of mind (this is, after all, a dynamic process requiring continual action) is dependent more than anything else on the state of being in which we learn to exist. And there are two general states of being available to us: the *mind state* and the *Self state.*

CHARACTERISTICS OF THE MIND STATE

When you are centered in the mind, there are certain behavior patterns that predominate and are unavoidable. These will be discussed later in detail. What is important at this point are the results of these behavior patterns. Living in the mind state always results in four general conditions: *fear, boredom, predictability,* and *dis-ease.* In fact, we can define the mind state itself in terms of these four qualities.

Fear

Individuals existing in the mind state are, first of all, afraid. They spend much of their time protecting themselves from all the evils of life. They worry about money, about their jobs, whether people like them, whether they'll succumb to some new disease, whether they will be mugged, whether their spouses are cheating on them, and whether the world will end tomorrow. Now that is admittedly an extreme example, but the fact remains that whenever a person has an excessive or irrational fear, it is the result of dwelling too much in his or her mind. Think about it. You probably know people (and for that matter have personally experienced times when it is true of you) who seem to be obsessed with protecting themselves. They are generally so busy trying to head off disaster that they miss the opportunity to live life in the process of protecting themselves. Everyone dwells in this state from time to time and to some degree. It is a natural consequence of the lifestyle of today's world, where uncertainty is everywhere, where change is ever more rapid and often traumatic, and where the news of the day is always bad. It's a wonder we're not all a bit more paranoid than we are. To paraphrase, if you can keep your head while those about you have lost theirs, you probably haven't grasped the seriousness of the situation. This is a fear state, and it is an absolute consequence of not understanding the dynamics of the world.

Boredom

The second characteristic of the mind state, *boredom,* is equally unavoidable if we choose to live our lives thoughtlessly. Since the mind state is forever attempting to head off possible disaster, it tends to minimize surprise and wants only the expected to happen. Without surprise and spontaneity, everything becomes preordained and therefore you become bored. People in the mind state prefer to know what is coming. They exist in a rut. A rut is nothing more than a hole that is open at both ends, and this is exactly where people living in the mind state find themselves. They find no joy in new events, since they seek to minimize them. They find themselves following social ritual, which has the advantage of utter uniformity and sameness while protecting the practitioner from the frightening prospect of something new. They structure their lives to eliminate every last visage of the unpredictable and often refuse to acknowledge it when it happens. As a result, people in the mind state are just plain bored. To whatever degree you are bored, you are existing in the mind state.

Compare the actions of a small child and that of the child's parents on a wet, rainy Saturday morning. For the child this is a wonderful event. Who cares if it's raining? The possibility of water falling from the sky is quite high, particularly in the summer. You can watch the water racing down the street, along the edges, pulling the flotsam of sticks and leaves and silt along with it. You can study the patterns of currents as they turn corners, come up against a barrier of collected trash, or carve new channels in the soft sand of a stream's bank. You can splash in it, feel it on your face, sail toy boats in it, or wash you bike without having to use a hose! You can go hang out in your tent in the back yard and pretend you're deep in the jungle in the rainy season or mount the deck of an imaginary ship in a great gale. The list of things to do on a good rainy day is endless!

Is this the way parents view a rainy Saturday morning? Probably not. For the mind-state oriented adult, a rainy Saturday means you can't get the grass cut or go to that baseball game. Instead, you're stuck at home, hoping the storm doesn't get too bad and disrupt the power or hoping the roof doesn't leak. Rainy days are gloomy inconveniences for people in the mind state. They are wet and dank and mean having to put on extra clothing just to step outside. They mean slick roads and delays. They are totally miserable events to be much avoided, and the tendency is to do just that. So you sit in front of the great cyclopean tube and watch the same shows you always do, or do the checkbook, or actually talk to your family. But whatever you do, one thing is for sure: You're not doing what you'd rather be doing, and you're bored! In fact, the boredom syndrome for the mind-state dweller is so pervasive that if it were sunny rather than raining, you'd probably be complaining about having to cut the grass or having to go sit in a hot stadium because the game is not being televised.

Predictability

All this leads us to the companion characteristic to boredom, *predictability*. With the mind-state oriented individual, predictability reigns supreme. Remember what the mind state does. It seeks to minimize surprise and to head off any possible negative events in a person's life. This necessitates sticking to behaviors and *modus operandi* that are proven to be safe and to create the same results every time. Social ritual is a very good example of this. Mind-state people have specific ways of doing things and will always behave in the same manner. Getting up in the morning has a routine to it that guarantees that all necessary activities are carried out in the same order, in the same manner, and in the same efficient style. There are no surprises. Monday is cereal day, Tuesday is breakfast downtown, Wednesday is scrambled eggs, and so on. Getting up becomes a matter of (1) brushing your teeth and shaving, (2) feeding the pets, (3) putting on the coffee, (4) showering while the coffee perks, (5) putting on the clothes that you laid out the night before, perhaps even the same outfit every Monday, another specific outfit on Tuesday, and so on. Everything is scheduled. Everything is programmed and preordained and expected, and there is little or no time for flexibility or variance. And what if something should go awry? What if there is an unexpected event, such as the power going off in the night or the alarm clock not going off at the right time in the morning? Well, that's okay with structured mind-set persons, because they usually have contingency plans already laid out to take care of those annoying unexpected occurrences that mess up their lives from time to time. Life becomes similar to playing chess: The great masters of the game study every possible move and combination of moves until they can predict four moves in advance how long it will take to checkmate the opponent. Future events become merely past events that happen in the future. And you can set your watch by them. You can always count on these types to do the same thing. You know their patterns and the predictable nature of their lives, and except for the amusement of watching them perform their act over and over, people who do not dwell in the mind state find them to be (of course) very boring. And they are. In contrast, do not think that there is no value in having routines or structure in your life. Because of the complex nature of life, some order and routine are necessary for all but the most fluid of people. The issue is not whether you have structure or not but whether you have *choice* about that structure. The key to understanding the behavior of people in the mind state is that *they have no perceived choice!* In their minds, things must be done this way in order for them to be happy and survive.

Dis-ease

Finally, we have *dis-ease,* the fourth characteristic of existing in the mind state. Notice the hyphen in the word; it is there because dis-ease is a state of being ill at ease. Thus dis-ease can be physical, spiritual, or mental. Any of the three aspects of being a person can be affected by dis-ease, and generally, if left unchecked, a dis-ease in one aspect of life bleeds eventually over into the others so that a truly mind-set dominated person ends up sick in mind, body, and soul. We are aware that fear can create illness of both a physical and mental nature. Fear states create extreme shifts in bodily response, including more rapid breathing, higher blood pressure, an adrenaline push as the body prepares for flight or fight, and a host of other physiological changes that put undue stress on our physical bodies.

The result through time can be everything from diabetes and hypertension to cancer and ulcers. Yet it goes far beyond this. Psychologists tell us that our attempts to adjust to the tensions and fears in our lives can create truly horrendous consequences for our mental state, manifesting in delusions, hypochondria, schizophrenia (which has both physical and mental components), intractability, aggressiveness, and the list goes on. Somewhere between 80 and 100 percent of all physical illness has a psychological base and is the result of our mind-state beliefs. Dis-ease is a function of the mind. If we do not take steps to avoid creating this dis-ease, it will predominate in our lives. The proof of this can be found all around us.

Look at the people whom you know to be predictable, repetitive, and protective of themselves. Look at those who express fears and take excessive precautions in their lives. Look at those who experience trauma and tension and bury it rather than deal with it. Now look at their state of health. This is the result of living in the mind state. It can be said that the cause of all dis-ease is nothing more than error thinking. It is the way we perceive the world and the way we think that creates our lack of ease with ourselves, with our environment, and with other people. In seeking to protect ourselves unreasonably, we sow the seeds of our own demise.

This is not a pretty picture. However, it doesn't mean that we should not pay attention to the world around us, or have order in our lives or routines, or deal with real problems as they arise. Such an extreme would be just as disastrous. What it does say, however, is that being excessively in the mind state to the exclusion of other possibilities is an imbalance that will eventually kill you and certainly make you miserable in the process. Fortunately, there is an alternative. Read on.

CHARACTERISTICS OF THE SELF STATE

The alternative to living in the mind state is living in the Self state (in this case, the word *self* is capitalized to delineate it from the ego-centered self). The self that we are talking about here is the true Self, the inner being free of attitudes, defense mechanisms, and the vast array of fears that permeate the average memory. We are talking about the Self in which our values dwell, the Self that operates on first principles. Just as with the mind state, the Self state has characteristics that define the behavior of someone living in that state. These characteristics include *health, love, wealth,* and *perfect self-expression.*

Health

In this context, *health* refers to living in a state of balance as opposed to the dis-ease of imbalance among physical, mental, and spiritual aspects of your life. Just as dis-ease can come about as a result of the turmoil and tension of life, health results from living life in a balanced, responsible manner in which you accept things as they are, move instinctually and appropriately to changes rather than attempting to avoid or stop them, and generally see life as a flowing of events rather than as challenges to be met or frustrations to be avoided. There is generally little difference in the actual content of lives, whether mind state or Self state, but the way in which they are viewed and met makes all the difference. It still rains, it still snows, roofs still leak, and work still has its problems (opportunities), but the way this content is viewed, that is, the context in which you hold these events, makes the difference between health and disease. There is an acceptance and confidence among those dwelling in Self that stems from the balanced, accepting attitude of the individual. Rather than being ill at ease with change and the need to react to the unknown, Self-oriented people merely get on with the process and wait to see what they can learn and gain from the events of their lives. They are more interested in solutions than in problems and tend to approach life in a more positive way.

There is an analogy that can be drawn here between the health of mind, body, and spirit and the three-legged stool of Freudian psychology involving the id, ego, and superego. In simplistic terms, according to Freud, to be mentally balanced, each of the three legs of the stool must be equally strong, no single aspect dominating and no single aspect being deficient. If they are not equal, the stool leans and eventually collapses. The same holds true for mind, body, and spirit. By balance, we mean that each is equally cared for and equally developed, resulting in a balanced and healthy life. If one aspect predominates, if you too strongly emphasize one aspect to the detriment of the other, then you have a sick body, mind, and spirit and, therefore, are not able to achieve peace of mind. This is not a new idea. Numerous religions, philosophies of life, and cultural traditions the world over express this same idea. In the ancient Celtic tradition, a primary symbol, the Celtic triple knot, stands for these same three aspects of life that we have been discussing—the mind, body, and spirit. In Christianity the commandment to love God with thy heart, mind, and soul represents the same triumvirate. Health requires balance, a subject about which we will learn much more in other chapters.

Love

The second characteristic of the Self-oriented (do not confuse this with self-centered) individual is *love.* Here we have to be careful in our understanding of the term, because love is an often misunderstood term, probably because it has so many meanings. In the sense it is used here, we are speaking not of physical love (eros) or brotherly love (fraternitas) but of universal love, or love of our fellow human beings (agape). In this sense, it is just a general feeling of affection and desire for the continued well-being of others that constitutes a feeling of love. And this is a much more natural process than most people think. Be aware that a desire for the continued well-being of others is the natural state rather than the exception. The truth is that we must be taught, through the process of socialization and experiencing the fears of others and ourselves, to not love others. In fact, fear is essentially the opposite of love. Usually, if you ask someone what the opposite of love is, he or she will say hate. But this is not the case. Hate is nothing more than love held in; it is love that is not shared but rather withheld because of some sort of fear. A person's capacity to hate and capacity to love are equal; in both cases we are dealing with the same amount of love. The question is simply whether that love is offered or held back. The true opposite of love is fear, and as we have seen, mind-oriented individuals are constantly reacting to and attempting to avoid the things that they fear. It is in the fear that the misery of life exists. An interesting truth is that as we move closer to a feeling of universal love, we experience lower levels of fear. One literally annihilates the other. And with the shift to a loving attitude from an attitude of fear comes increased contentment, increased balance, and increased peace of mind.

Wealth

Wealth is the third characteristic of the Self state. In this volume, the word *wealth* does not refer to large checking accounts, to huge houses or penthouses in some major city, or to vast tracts of country land; it has a much broader application than merely physical possessions and amassed fortunes. In the context in which we use it here, wealth refers to abundance in all things, not just some.

For each individual, there is a different definition of exactly what wealth means. For some, it is indeed great physical wealth; for others, it may merely be a matter of never wanting for anything. Some people find themselves wealthy when they are able to pursue their chosen careers and have the opportunity to practice their profession; others find abundance in the natural world around them. No one of these definitions of wealth is superior or inferior to another. Wealth is a matter of personal definition. For that reason, it is more accurate to speak of abundance, which is the opposite of need.

In the case of abundance, whatever a person needs is available to him or her. Whatever is necessary for the well-being and success of the individual is extant and in the possession of the individual when it is needed, and it is enjoyed without guilt, without remorse, and without fear of loss. That is abundance, and that is what living in the state of Self really means. As we will see, one of the benefits of an ethical life is the capacity to create abundance whenever and wherever it is appropriate and, often, at will. At this point, that may sound ridiculous or impractical, yet as we shall see, this is far from the case.

Perfect Self-expression

Finally, we have *perfect self-expression.* This is a primary characteristic of the Self state. Perfect self-expression refers to the capacity to openly and honestly express who and what we really are, without the fear of how it will be viewed and without ulterior motives. It is what Abraham Maslow calls self-actualization, or the ability to be yourself. It is what we refer to when we think of someone as being natural and without guile, or what we observe in small children who have not yet been taught to not be themselves. There is no attempt to control others or defend one's beliefs when you experience perfect self-expression. One of the main benefits of this type of behavior is that you always act appropriately in every situation, and you do it naturally, without having to think about it or decide which road to take. It simply happens. Human beings are at their best when they are expressing themselves perfectly. They are, in fact, offering the only gift they have to offer humanity—that is, themselves. Every moment becomes an event unto itself, an opportunity to experience and do. There is no opportunity for boredom in the ever-changing world. There is no need for predictability; whatever is appropriate automatically presents itself at every moment in time. With perfect self-expression, there is perfect health because there is no error thinking; there is always a very natural and very real feeling of affection and love for others. There is no motivation to feel anything else. Almost magically, you become perfectly ethical; dwelling in a state of Self essentially means doing what works.

Yet the question remains, What is the overall result of this living in Self? The answer is simply that we have everything we could possibly want, and we have it without fear. We know that what we need is there for us and that it will be there for us whenever we wish, in a form that is most appropriate for our well-being. In other words, we have *peace of mind.*

PEACE OF MIND

As stated previously, for the most part, when people think of happiness, they really think of peace of mind. So are they not the same? In a sense they are, yet as the word is used here, it refers to a specific component of peace of mind centered in contentment. You can be troubled by conditions or saddened by the loss of someone in your life and still be content to let events unfold. Contentment comes from the knowledge that things happen as they are supposed to and that there is nothing to fear (there it is again) if you just relax and move steadily in the directions you wish to go. In general, achieving this state is a matter of experience; once individuals open themselves to the possibility that the events of their lives are for their benefit, no matter how they appear at the moment, the truth of life's benevolence becomes more and more obvious. That's a radical statement and not one that I would expect a reader to accept immediately. In truth, there is no need to accept it at all. A great deal more will be said about this later, and it is hoped that the efficacy of the statement becomes more obvious as we go along. Suffice it to say that the contentment experienced in the Self state stems from this realization that everything is truly okay and that it is simply because of an individual's inability to see the larger picture that it seems otherwise.

By way of example, consider the case of an individual who loses his or her job. For a person in the mind state, this is a catastrophic event. "I have bills to pay! I have a family to support! My gosh! I may never find another job as good as this one! What am I going to do?" Immediately the mind-state person begins to see the dangers of the situation, to panic in the face of a major blow to his or her routine, and to experience extreme mental anguish, often followed by depression and even physical pain. For the Self-oriented person, the response is much different. It is not a catastrophe but simply an event. Things have changed. New behavior patterns must be developed and actions taken in the face of the event. It is not necessarily a terrible thing. Even though there are major changes in life and lifestyle, the Self-oriented person is content to deal with them, realizing that this is just an event, perhaps one that they would prefer had not happened. Although they may be displeased by it, they have the wisdom to be content in knowing that overall this is just another step in the process. Perhaps it will lead to an even better job. Perhaps the company for which they were working is experiencing serious difficulties, and when it collapses later, this individual, by virtue of having been fired, will not go down with them. There are new avenues to explore and new possibilities to investigate. Rather than dwelling on the losses of the past, the Self-oriented person dwells on the future and what to do from here. In both reactions, the seed of content or discontent stems from the way the event is viewed, not the event itself. The key here is the *context* in which you hold the event, not its *content*.

COUNTERPOINT AND APPLICATION

At one level, being ethical may seem to be counterintuitive; we constantly see people around us who act in ways that are considered unethical, and they appear to be getting away with it—from the common thief who hasn't gotten caught to the office politician who gets ahead through manipulation and intrigue rather than doing a job well, to the major corporation that lies to the public and controls market functions to its own advantage at the expense of others. If all this is true, why should we bother to lead an ethical life? Quite simply, it is because the seemingly successful people performing unethical acts are not really getting away with anything. No one ever does.

I remember giving a test in an ethics class one term and having a number of people cheat on that test. I found it ironic that they would cheat on a test in a class on ethics and decided that I must not be doing my job well if they hadn't learned enough not to cheat. In actuality, two processes were going on. First of all, I left the class during the test, preferring to treat these people as mature, responsible adults rather than recalcitrant children. It was an action of respect. Secondly, when asked why they cheated, they laughed and informed me that I had said they should do what worked, and cheating worked to get them a good grade. Their smiles vanished when I pointed out the following: (1) they had made zeros on the test and indeed failed the class, so the process did not work as they thought it would; and (2) if their only purpose was to get a good grade on the test rather than learn the material, they were lying both to themselves and to me as to their motivations for being in the class in the first place.

The point is that unethical activities never work. Eventually, those behaving in unethical ways are caught or at a minimum experience a wealth of negative consequences as a result of their behavior. Doing what is expedient is not necessarily doing what works. As a test, consider the following question: If you were in business, would you want to do business with yourself? Now there's a sobering thought. How ethical are you in your dealings with others? Would you want to have business dealings with someone like yourself?

In terms of technology, the same applies. What are your motivations for creating or supporting a given technology in the way that you do, and would it be okay with you if someone else operated that same way? If you were in a position to run everyone else out of the market because of a new technological development you controlled, would you want someone else to do that to you? How do you implement the new technology in a manner that benefits you while not operating to the detriment of others? If you are thinking that what happens to others is not your concern and that it is not your job to protect others from the effects of technological change, how does that mesh with our definition? Before I confuse the issue too greatly, I will offer this hint: It is not your responsibility to protect others from the effects of progress. It is also true that what someone sees as being detrimental to him or her may in fact be a benefit in disguise. Have you ever lost a job only to find it opened up new possibilities you never would have considered otherwise? Perhaps adversity is a matter of perspective.

We choose to dwell either in the mind state or the Self state. In the first case, we deal with fears and problems caused mostly by our own unconscious understanding of our behavior. In the second case, we know that our actions and behaviors work for us and that even seemingly negative events have a positive effect on our lives, if we will but use them to gain wisdom and trust in the process. Solutions present themselves magically if we dwell in a state of Self. Seeming roadblocks become new pathways or magically melt away. We move effortlessly from event to event, experiencing life and participating in it rather than seeking to avoid it. We find ourselves healthier, more peaceful, excited about life and free to express ourselves in creative and constructive ways. These are the reasons for behaving ethically, each one practical and desirable, each one obtainable.

EXERCISES

1. To develop a sense of the degree to which you live your life in the mind state, write out your normal routine for the day. Pick a typical weekday for this exercise and make a detailed account

of what you do. Now look at the elements of this list of activities. How many activities do you perform invariably and how many most of the time? Now think about why you do them. Are there some that have become so routine that you would be extremely uncomfortable not doing them for any reason? Are there alternatives to how you perform your activities?

2. Now create a different activity with the same purpose as that listed in exercise 1. Do you always know what to wear? Do you always arrive at work or class five minutes early? Perhaps you always wear the same tie or belt or blouse with the same outfit. Try some appropriate variations and see how you feel. You should get some idea of how dependent you are on your routine. What elements of your old routine have you flatly refused to change? Write a few paragraphs on why.
3. On three-by-five-inch index cards or in a small notebook, begin to jot down the times when you feel bored. What is going on at those times? As best you can, describe what it feels like to be bored and what it is about the current situation that creates the boredom. Incidentally, notice how this process of paying attention to your behavior relieves the boredom. After you have done this for a few days, look back over the list and search for patterns. If we assume that the boredom stems from your own interpretation of the events you find boring, how can you see them in a new light to change your boredom to some other state? Be creative.
4. How predictable are the people with whom you are familiar? Observe how often you know what they will probably say and do under a given set of circumstances. How true is this of you as well? Now speculate, or even better, ask them, why they always do things in the same way. Explore the reasons for your own predictability as well. Again, change those things about you that are predictable. Pick a few and simply become more spontaneous. Note the results in your notebook.

CHAPTER 4

TECHNOLOGY AND THE SELF STATE

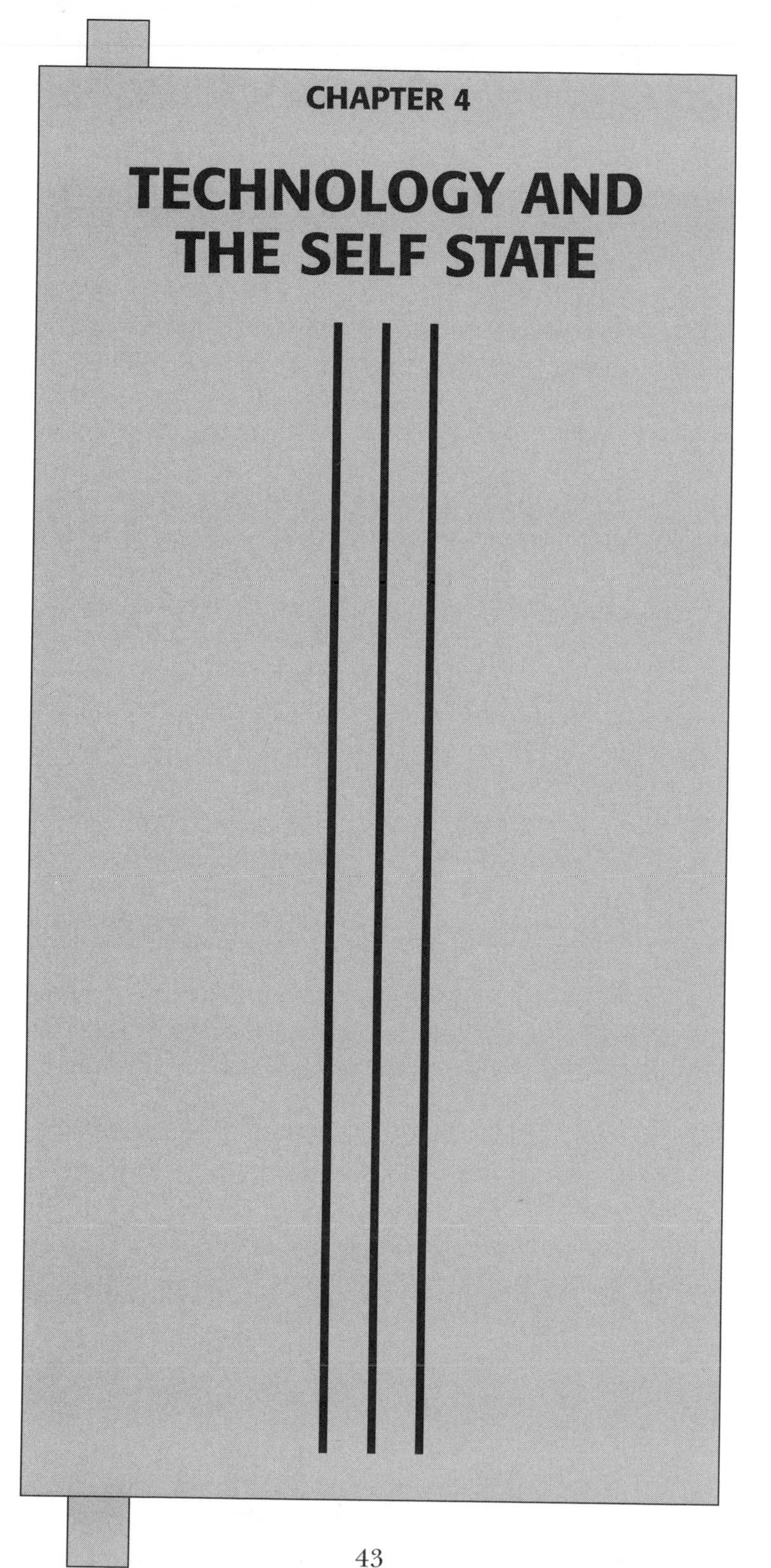

INTRODUCTION

It is important to not confuse the mind state with intellectual action. Without intellectual action, there could be no technology to begin with. In fact, the intellectual process of creating technology is very natural and very much in tune with being yourself. If we look at the species we call Homo sapiens from a realistic "scientific" point of view, we find that creating technology is purely an expression of individual freedom and self-expression.

A REMINDER ABOUT TECHNOLOGY AND MAN

As mentioned earlier, we have received two very powerful survival mechanisms from our natural evolutionary process: the ability to produce technology (technologizing) and the tendency toward homeostasis (resistance to change). These are practical mechanisms and highly steeped in normal intellectual processes. Yet by virtue of their natural character, they involve not only the characteristics surrounding problem solving but also the process of determining and finding solutions. This solution element of our survival is not a mind-state proposition. It is of the Self and is therefore centered in the process of living in Self. Hence, we have a situation in which we naturally discover new and unique ways to provide ourselves with what we need for survival in a changing world and then resist those new ways in a homeostatic desire to protect ourselves from the unknown consequences of those changes.

We view new ideas as "foreign," or strange. We look upon new gizmos as unfamiliar and suspect devices to be considered before adoption. The more radical a departure from the norm the idea is, the more difficult it is to achieve acceptance in the society. In this way, we can rapidly create changes and, at the same time, absorb them slowly enough to avoid major disasters in the process. There are going to be errors, as in the case discussed earlier of traditional agricultural settlements destroying themselves through their rapid acceptance of new technology. Many other examples can be found as well. But in each example, the effect was relatively local, and because of the degree of suspicion and isolation among tribal groups or larger cultures, there was still time to determine the value of a cultural process before it was fully incorporated. The anthropologist distinguishes between the process of *acculturation* and the process of *assimilation* through which this cross-cultural adoption of ideas can take place. In the case of acculturation, two cultures that come into contact with each other for an extended period of time will often find themselves adopting ideas and technologies from each other as they find them useful. Hence we see American cuisine incorporating dishes from Asia, South America, the Indian subcontinent, and the South Pacific into its generally European-based approach to cooking as North Americans come in contact with the peoples of these regions who have moved for various reasons to the United States or Canada. Similarly, traditionally European cuisine enters the lives of those living in Mexico or those who have traditional cultures within the bounds of the United States and Canada, such as Native American cultures. We should distinguish here between *acculturation,* which occurs through direct contact, and *diffusion*, which can occur without that contact. The acceptance of Coca-Cola worldwide, for instance, or the adaptation of Western dress in Asian cultures may not depend on actual contact so much as awareness and borrowing.

On the other hand, *assimilation* is a much more extreme adoption of cultural and technological ways in which one culture is absorbed entirely by another larger one, though the larger may take on some of the characteristic elements of the smaller culture in the process. In either case, there is initial suspicion of the foreign ways of each culture by the other, and in the case of acculturation, this allows time to observe a new technology before it is accepted.

We find further complications of this process in the twentieth century and beyond. That complication stems from the increased contact among cultures through rapid, inexpensive travel and, in the last twenty years of the twentieth century, through the explosive increase in communications technology. Suddenly it was possible for a new idea to be presented to a world environment through radio, television, and the World Wide Web. Such rapid dissemination of information greatly reduces the time span between initial discovery and potential general adoption of a new idea or device. Simultaneously, the rate at which changes take place and the degree to which people react to those changes with fear also increases. Historically, too rapid a rate of change has resulted in social upheaval and even revolution. We simply cannot assimilate too much change too quickly, and as our ability to create that change rises disproportionately to our capacity to handle the changes, the culture suffers.

This being the case, where is the relationship between all of this theory of technology and the subject of technological ethics? Simply put, *it is in our natural capacity to create technological change that we find our greatest opportunities and our greatest threats to survival.* This would seem to indicate that the creation of technology is innately good, in that it utilizes exactly the survival traits that nature has supplied for making our way in the world. It would be nice if life were that easy, but unfortunately, there's a bit more to it than that.

Thus, it would seem a miracle that any new changes are actually instigated, though instigated they are. Human beings find change both exciting and exasperating, but the process of opposition between the traditional and the new creates a give-and-take that works quite well for us. The question concerning us now is where these new ideas come from.

What may be a surprise to many is that these ideas do not come from that part of the mind that deals with problems. They come instead from that part of the mind that deals only in finding solutions—that is, the unconscious. It is from the unconscious that we receive flashes of inspiration, where dreams are born and where the sudden insights that seem to spring full-fledged into our consciousness originate. We have no volitional control over this part of our mind. It acts without direct intervention and can operate at lightning speed compared to the conscious mind, which laboriously explores circumstances one step at a time. It is, in fact, where most of the mental work of our lives takes place. Overwhelmingly, the solutions to our problems come upon us as a result of a thought process that, though it feeds on facts and knowledge, is for the most part intuitive. How often have you gone to bed with a problem on your mind and awakened the next day with a solution? Is this a logical process or an intuitive one? If it is totally logical, why didn't you think of the answer before going to bed? How often have you looked at a situation and simply understood that given the way things were constructed, they would not work or, alternatively, that if certain changes were made, it would work better? You may not even know how to explain your reasoning, but you just know! That's the creative mind at work, and in order to work, it must operate not from a mind state but from a state of Self, a state of totally being who and what you are, naturally and without conscious thought.

Just as there are four aspects associated with the mind state, there are also four aspects associated with the process of dwelling in the Self state. These four aspects—wealth, health, love, and perfect self-expression—have been discussed previously. However, it is important to remember that achieving them requires that we be who we are, and that means creating technology. Doing what works is a matter of being efficient, and Homo sapiens most effectively does that through the use of technology. It logically follows from this that technologizing is a necessary part of being ethical.

TECHNOLOGIZING AND ETHICS

Just because we have the capability to create a technology does not mean that it is necessarily a good idea to do so. This is the reason for our homeostatic tendencies to resist change. Without that natural suspicion of the new, we would simply embrace every new idea and incorporate it into our lives and our culture, possibly with disastrous results, as mentioned previously. However, what we can infer is not so much whether a given technology is or is not ethical but whether the process of technologizing is ethical, and the answer is most certainly yes. We are not saying that it is ethical to produce all technology but rather that the act of producing technology in general is ethical, though the individual technologies developed and the manner in which we choose to use them may or may not work for our survival. This is not nearly so fine a nuance of difference as it may at first appear to be.

With a given technology, it is necessary to determine the impact of that technology and the purpose of that technology in addressing the degree to which it represents an ethical or unethical artifact. The Self state would necessarily includes the creation of technology. It is, after all, a matter of expressing who you are rather than denying it. To achieve health, wealth, love, and peace of mind, it is necessary for humans to be creative, and that means, at some level, developing technology. Beyond this issue of whether or not to technologize, it is necessary to investigate other facets of a given situation to determine its ethical efficacy. This is not a mindless process, merely a highly intuitive one. The capacity to produce something does not necessarily indicate that it is appropriate to produce. Indeed, the average human being on the street, with very little study, is capable of producing very dangerous and deadly technologies. We will speak more on this matter later. For the moment, in order to place things in a meaningful context, we need to look at some other aspects of what it is to be ethical, specifically, the role of reality states in determining what works.

COUNTERPOINT AND APPLICATION

All this sounds very agreeable, but it seems highly impractical and highly suspect. For an engineer or scientist, the approach is to work with the physical world and its content to determine how best to achieve some end efficiently, and/or to learn new information about the functioning of that universe. Dwelling in a state of Self and looking to intuition seems very unrealistic and impractical. Isn't that so?

Is it? Rene Descartes is said to have invented the Cartesian graph system while daydreaming and staring at the corner of his room where two walls and the ceiling came together. He suddenly saw them as three intersecting planes. Einstein developed his theory of relativity while dreaming of riding in an elevator. Edison invented devices by arranging to be awakened suddenly while in the process of entering sleep and then writing down whatever he was dreaming/thinking about at the time. These are all intuitive processes taking place in the subconscious mind, brought to the fore by analogy or sudden recognition.

No one is saying to forget all of the practical, hard core science and engineering. That would be disastrous and foolish. What is being said is that there is an intuitive, creative element that is part of the Self state, a part that works on solutions rather than problems (as does the mind state) and that is valuable in the process of creating technology. It is an integral part of our self-expression and thus absolutely essential to creating technology ethically. Engineering is an application of our understanding of physical laws, but those physical laws, no matter how often empirically proven, originated as intuitive flashes, often through analogy, which is the province of our perfect self-expression. To deny the Self state is to set ourselves up for failure or, at a minimum, to limit our success.

EXERCISES

1. List five technologies that do not exist yet are possible and that you consider to be unethical in that they would not support the survival of the species. For each of these five technologies, explain why you find them unethical.
2. List five technologies that do not exist that are possible and that you do consider ethical in that they would enhance and support the survival of the species. As for the first question, explain for each of these five technologies why you feel that its creation is an ethical choice.
3. Now consider why each of these technologies does not already exist. Is the consideration one of ethics or is it some other factor, such as profit, national prestige, political exigency, or cost?
4. Now look at the following list of technologies that do exist and decide whether they are ethical or unethical and why.

   a. Automobiles
   e. Rubber bullets
   b. Electric power
   f. Nerve gas

c. Computers	**g.** Gas masks
d. Apache helicopters	**h.** Television

5. In those moments when you feel as if you are living in a Self state (that is, you have peace of mind), what are you doing? What kind of activities are you involved in? What type of creative thinking are you using? What does this say about how you live your life?

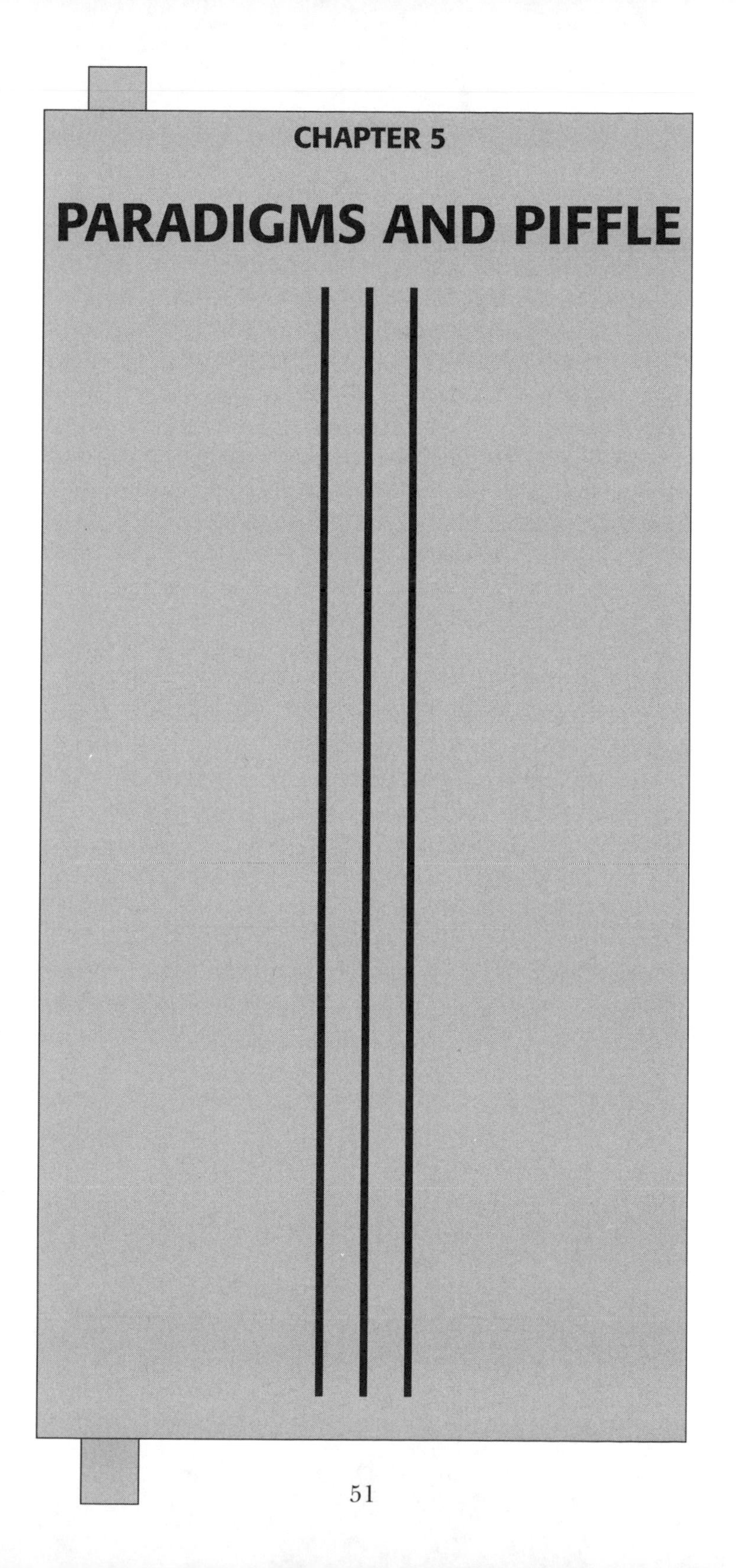

CHAPTER 5

PARADIGMS AND PIFFLE

INTRODUCTION

Paradigms are nothing more than belief systems. Indeed, Thomas Kuhn describes a paradigm as "universally recognized scientific achievements that for a time provide model problems and solutions to a community of practitioners" (Harris 1979, 19). Clearly, this is a technical definition, born of the necessity of understanding scientific processes in the light of current belief systems. And Kuhn firmly believes in the necessity of this tendency toward paradigmatic centrism and views paradigms in opposition as a necessity for scientific growth and progress. Yet as we shall see, paradigms are not limited to the scientific community or even to an engineering/technological orientation. They are an integral part of all our lives as we strive to create our personal sense of reality.

PARADIGMS AND ETHICS: A FABLE

It was a beautiful spring day, and Jennifer, Jane, and Joanne were walking along a shady downtown avenue after their usual Wednesday lunch together. Two of them were discussing the new boss at the company where they worked together while the third, Joanne, was listening with only half an ear. Her attention was more centered on how best to deal with her son's recent resistance to doing his chores. It was a typical after-lunch walk for all of them, made more pleasant by the warmth of the sun overhead and the gentle breeze rustling with a hiss through the leaves.

A woman and a small child, both dressed in casual clothes and holding hands, were coming toward them. Suddenly, the woman turned to the child, yanked him by the arm, and spanked him hard across the bottom. The sound of the impact could be heard clearly from where the three were walking, half a block away, and the force lifted the child momentarily off the pavement. Seizing the child even more tightly, the woman opened the door of the car beside them and pushed the child into the front seat. She quickly walked around the car, entered on the driver's side, and started the engine. The child could be plainly seen crying as they drove off.

"Did you see that?" said Jane.

"I certainly did!" answered Joanne. "I think we should call the police! Did either of you get the license number?"

Jennifer merely looked at them quizzically. "What are you two so excited about? Why call the police?"

> Joanne looked at Jennifer in amazement. "Didn't you see that kidnapping that just took place? Weren't you looking?"
>
> Jane shook her head. "Now Joanne, I don't think it's *that* bad. After all, just because that mother was abusing her child, we can't know for sure that she was kidnapping him. Do you think the father has custody? Was that an abduction?"
>
> Jennifer began to laugh. "I don't know what you two think you just saw, but all that happened is that an unruly child was firmly disciplined by his loving mother, though I must admit she seemed very aggravated. I always thought it was a loving thing to do to teach a child respect and discipline, even if it does require an occasional swat on the bottom."

So what happened? What did they really see? Was this a kidnapping? Was this a matter of parental abduction? Was it child abuse, or was it just parenting? Do any of them really know? In this scenario, each of the three women had her own version of what happened. Each saw the same exact set of actions, but each interpreted what was happening in an entirely different way. In fact, they experienced three entirely different events.

You have probably experienced this type of multiple interpretation in your own life. In fact, it is virtually impossible to skip through modern life without this happening, either occasionally or often. Why? What has caused such totally separate and distinct interpretations of the same behavior?

The answer lies in the *paradigms* of the three women involved. A paradigm is a worldview, the individual's view of how the world is put together and how it operates. We all have different paradigms, based on our experiences, beliefs, and the things we were taught as we were growing up. In fact, one's paradigm is the result of the sum total of all events of a person's life and the manner in which the individual interprets those events. It creates for each of us what is essentially a separate and distinct reality.

Just as the three women developed a different scenario of what had happened in the incident, based on their own worldviews, so do each of us view life in accordance with our understanding of it. It does not mean that we are right in our assumptions, nor does it mean that we are wrong. *It merely means that we each have a separate reality in which we live!* We literally live in separate worlds.

There is an ancient fable about three blind men who are asked to describe an elephant. One of the three reaches out and examines the animal's leg with his hands. The second reaches out and comes in contact with the elephant's body. The third encounters the trunk. The first blind man says, "An elephant is like a great tree trunk with branches too high to reach." The second insists, "You are wrong. It is a great leathery wall." The third opines, "How can you both be so wrong? An elephant is snakelike and breathes loudly as it wraps itself about your limbs." Obviously, each has described some of the characteristics of an elephant. Yet no one of them is completely correct; each has encountered only a small part of what an elephant is. Probably if questioned further, they would all find something to agree on about the nature of elephants. If you have ever been close to an elephant, you would probably agree on the nature of its odor!—that is, unless you were experiencing a stuffed up nose. Which brings us to a primary issue in being ethical. How do you know what is really true and therefore a basis for making ethical decisions?

Earlier it was suggested that the nature of ethics might be culturally bound, that what is and is not ethical depends on the culture from which people come and the conditions under which they have been raised and are living. This is partially because circumstances create a personal paradigm. In one culture, eating red meat is acceptable; in another, killing a sacred cow is the ultimate in blasphemy. One culture views cannibalism as shameful and absolutely taboo; another dines ceremonially on the flesh of departed relatives. Who is being ethical and who is not? It depends on the individual's paradigm and upbringing.

Does this mean that there are no specifically ethical principles? Are they all relative to the environment and the culture? People have been arguing this for thousands of years and will probably do so for thousands of years to come. The reality of ethics is argued even within a given culture. Stealing is considered unethical (not to mention immoral and illegal) in nearly all cultures. Yet is stealing a loaf of bread to save a starving family an unethical act? Isn't that a matter of condition? How do you decide? This is the problem with relative ethics, and it is brought about because we have different senses of reality.

Yet there is a solution to this dilemma. It lies in understanding the difference between paradigmatic reality and true reality, and that is the subject of this chapter. The question becomes, Are there concrete immutable principles of ethics, or is it all a matter of opinion? Just as people have tackled this question for thousands of years, so will it now be tackled here.

PARADIGMATIC ROOTS

To begin the process, let's take a closer look at the creation of our individual paradigms. Be aware that whatever your paradigm, you came by it "honestly." That is, it was the result of honest attempts on your part to make sense of this world. The information we gather early in life has a dramatic effect on our attitudes and beliefs in later life, both consciously and unconsciously. To categorize, we can break this learning process into those things we are taught, those things we learn by example, and those things we learn by observation and experience. An examination of each will provide a better feel for what is involved.

Those Things We Are Taught

When children are born, they arrive with instincts that allow them to survive long enough to learn how to function in the world. These innate understandings of the world basically dictate our desire for nourishment, warmth, and physical contact with others and (yes, it's true) our fear of falling. But this only serves to get us started. What happens next is a process by which our parents, our siblings, our friends and acquaintances, and the culture in general teach us what is and is not successful behavior in the opinion of that society. From earliest childhood, those around us instill the rules and regulations of the society, and when we behave in unacceptable ways, we are quickly corrected. The part of the process to be discussed here is the official part, where instruction is offered.

Imagine a small child out playing in the yard. Children are naturally curious, this being an integral part of what it is to be human. They are inquisitive, unafraid for the most part, and audacious. To them, a backyard is a paradise. A tree or a fence is a wonder to be explored. A blade of grass, an ant hill, or an insect is the object of passing fancy or scrutiny, depending on the mood of the child at that moment in time. Under such circumstances, it is normal to expect children to climb the tree, follow the movements of the insect, or precariously balance themselves on the top edge of the fence. Think back to your own childhood. How often have you walked the crest of a fence or followed the edge of a sidewalk to discover how far you could go before losing your balance? So imagine a curious child walking the crest of a wooden fence, carefully balancing the forces acting upon herself, seeing herself as a tightrope walker, high above an adoring crowd. What a fun experience! What a joy!

And then Mom comes to the back door and sees her child in action. "Georgia," she cries, "get down from there before you fall and break your neck!" Immediately, two things happen. Georgia, unless she has learned to not believe everything her mother tells her, immediately loses her balance and topples to the ground. Secondly, she learns that she cannot balance on a narrow surface and must stay away from such places. Is it true? It is for Georgia. Here is a normal, fully functional person with the capacity to do incredible acts of balance with practice and development, yet she probably never will. Next door, a small girl named Mary has a similar experience with the exception of her mother's reaction, which is "Do be careful, darling. I wouldn't want you to fall." She also learns. She learns first of all that she needs to be careful in this environment and secondly that if she is not careful, she may fall. Which of these two will continue to walk fences? Which will be willing to take risks but be careful? Which is convinced that if she falls, she will break her neck? Which will be willing to entertain the possibility of becoming a gymnast and walking the balance beam? The chances are that we will never see Georgia in the Olympics. But what about Mary?

At first inspection, these differences in perception may appear to be minor, but the truth is, they begin a pattern that could last a lifetime. A single event in each of these girls' lives has sown the seeds of separate paradigms for a small piece of their reality. This is a simple illustration, but the point is profound. How much of your beliefs about the way the world works and your understanding of right and wrong is the result of what you've been told from early childhood?

We must be careful here to acknowledge that there is nothing inherently dysfunctional about being taught how to behave in this world. Indeed, if we did everything by trial and error, many of us would never make it to our majority. Societies instruct their members in what works for the community to create order, provide the necessary means of survival, and live in harmony and survive as long as possible, and each culture has its own ways of doing that, based on the collective experience of its members. This is valuable information, passed on from one generation to another. The only differences in the process are the means available for the process and the complexity of the culture, which dictate the complexity of the material to be learned. The process is essentially the same, though the methodology and difficulty in learning "the rules" of a society change with such factors as population size, communications, and technology. It is a primary characteristic of culture to pass on the learned joint information of that culture through the generations. Yet what is it that we learn along the way? What is it that we internalize as our version of reality? One culture may teach cooperation; another may teach competition. One culture may emphasize hunting and aggression as a method of survival; another sees the world in terms of agriculture and accommodation. Again, the more complex the society, the more difficult it is to determine exactly what the successful rules for behavior are, and the truth is, it keeps changing.

I am aware of a young man, a brilliant individual, who was a very successful civil engineer. He worked for a major transportation company, and as a troubleshooter, it was his job to determine the cause of accidents when they occurred. He worked his way up to the position rapidly after joining the firm and was the youngest person to hold each of the positions he had in that company. And he was very good at what he did. His reputation as an analytical engineer was widespread. Yet he was not really fulfilling his true potential. Initially, if you asked him why he had become an engineer in the first place, he probably would have answered that he never really thought of doing anything else. It was just what he was supposed to do. But eventually, he began to ask himself why. It was not that exciting for him. He did not feel naturally drawn to it. Yet here he was, functioning as an engineer. After a great deal of introspection and self-analysis, he finally traced his desire to become an engineer to a single event of his childhood. He had been in the backyard, playing in a stream, building a mud dam, and watching the way the dammed-up water found new routes around the barrier. As each one appeared, he would reconstruct the dam to account for the new path of the stream, until he had developed a rather elaborate series of levies and traces. As it happened, his mother and a friend were in the yard at the time, and having noticed his behavior, the friend said to his mother, "You know, I think your son would make a wonderful engineer." It was just a passing remark, an idle comment by a neighbor that stuck in his mind and eventually led to a career. What we are taught, consciously and unconsciously, has a tremendous effect on us all, and the beliefs that we develop both consciously and unconsciously determine our view of reality.

So we busily go about the process of teaching our children how to behave ("Don't put your finger in that light socket. It could kill you!" "Finish your homework or you won't be able to go to college!" "Come to me when I call you!" "Of course you're going to church. Do you want to end up in hell?" "Don't play with those kids. They're not our kind of people!" "Clean your plate or you don't get dessert!" "It's a dog-eat-dog world out there, kid. You've got to be careful and you've got to be better than the other guy if you want to succeed!") and how to get along in the world, for their own well-being and their own benefit. Yet as we do, how much fear, distrust, aggression, and lack of confidence do we instill in them? How much do we force them into paths that we think are right for them based on our own paradigms and our own expectations rather than their strengths and their inclinations? How many of us are in careers that are the result of what we were told we should and had to do rather than what we really want to do?

As a final example, I offer the case of someone I know quite well. This person teaches where I teach and does it well. His story as told to me is as follows:

> When I was young, there was no doubt in my mind what I would do when I grew up. I would be a businessman, just like my father. He was good at what he did, very good. He had a very high degree of success. Beginning as a salesman in the 1920s, after not even graduating from high school, he had developed into the number one marketing representative in his company, then began his own company, which he successfully operated until he was in his late sixties. He had always shared his stories with me, of life on the road, how to operate successfully in business, and how, when I grew up, I would take over the firm and run it for my children and grandchildren as well. My education was in the field of business. I was always learning about business and how a business ran, particularly the family business, where I worked summers from the time I was fourteen until I was out of graduate school. I knew the business well, from manufacturing to selling to bookkeeping. And after graduate school, I held positions working in finance and marketing. Yet somewhere in the back of my mind, I was always remembering how much fun those two terms of teaching while I was in graduate school had been. Taking the job as a lark, I began teaching management as a way to prepare myself for the dreaded oral examinations coming up after my second year of graduate study. Surprisingly, it was the most fun I had ever had in a job. I relished every minute of it, but, of course, it was only a temporary respite.

> "Son," my father told me more than once, "I don't really care what you pick as a career [yeah, right, Dad] as long as you don't teach. Teachers are the genteel poor. You'll never make any money doing it. Any other profession is okay. Just don't become a teacher."
>
> Naturally, I ended up running the family business, and it was seven years before I realized that maybe my idea of life was not working. Maybe being a businessperson was not the be-all and end-all of life. Here I was, successful, making money, running a thriving business and doing it well, and I was totally miserable. The only joy in my life, outside of my wife and children, was the courses I taught part-time at a local junior college. Eventually, I got the message, gave up the business, and began teaching full-time. There have been no regrets.

We learn who we are and how to live from our parents, our teachers, and our society, and when it is done formally, we call it education. It is valuable, and it is necessary if we are going to be a success. Yet we need to be aware that in the process, it serves to create, from a very young age, our perceptions of how the world is put together and what works and does not work. How accurate are those teachings, and more importantly, what do we do with the information when we receive it?

Those Things We Learn by Example

The second way in which our paradigms are created involves the examples that others present to us, not in what they tell us but in their behavior and through their unconscious interactions with us. Children are not stupid. They are amazingly perceptive, and at a young age, they operate in a wide-open mode, in which they listen to everything that is told to them and that they observe. Sometimes this creates conflict, when the words spoken and the actions taken are in conflict. This is where we first learn distrust and where we become jaded to the content of the world. Again, in earliest childhood, we recognize the inconsistencies exhibited by people in their dealings with us and with each other. We encounter concepts such as the little white lie and the difference between what people do and what they want others to see them as doing. Have you ever heard a parent or other authority figure say, "If you do that one more time, you're in trouble!" Suppose you do "that" again? What happens? Are you spanked? Are you sent to your room or punished in some other way? Probably not. In fact, children learn early on that what their parents say and what they mean are not always the same. "Johnny, if you don't clean your room, I'm going to spank you!" Being a child, whose job is to test the boundaries of what you are allowed to do, you naturally see how long you can put off the job. "Johnny," you soon hear, "I really mean it. You'd better clean that room!" "Sure. I've heard that before," you say to yourself, and off you go about your business. "Johnny, I'm really getting angry now," you hear her say, and she has that tone in her voice that says she's losing patience. But you're still good for a little more fun and sun, so you just continue playing in the yard. "Johnny!" you finally hear, "Now that's it, young man. You get up to your room or I'm going to give you a spanking you'll never forget!" Now she's serious. You rush past her, up the stairs, and begin the arduous job of putting away your clothes or trying to figure out how to hide all those toys and found treasures with the minimum of work. So what have you learned? First of all, you've learned that when your parents say something, they don't always mean it and that you can tell the difference with a little testing. Secondly, you've learned that you can put off things for a considerable time without repercussions, even if they are threatened. The result is a child who views his or her parents as periodic liars and has thus learned to procrastinate. Was that the intended lesson from Mom or Dad? Their actions speak much louder than words. And these are the primary educators that first create the foundation of our view of the world. Later in life this could lead to the idea that it's okay to lie, that exaggeration is a normal way to communicate, and that nothing has to be done on time. All of these are examples of low integrity, situations in which people do not mean what they say or say what they mean. In terms of ethics, it's okay to not keep your word and to lie. And this can become a paradigm for the person who learns it. Why do we wonder at the deception in the world? Why are we becoming blasé about the ease with which people break contracts, falsely advertise, tell lies about situations and people, or go through divorce after divorce, each time reminding themselves that they didn't really mean it when they said those wedding vows?

Imagine a child of three or four running helter-skelter through a house. It's glorious to be this mobile. It's wonderful to move and run and use up all that energy, to feel the newfound coordination in one's body and explore it. And it's also not surprising that a child periodically will bump into things as he or she tests the limits of his or her abilities. So down comes a lamp or over goes a table, and along with the laughter or groans of parents and siblings, the child hears, "Good Lord, Sarah. You're so clumsy!" She hears that she is clumsy, and even if she is not, she may find herself learning to be clumsy. Consciously she perceives that she's in trouble and has done something clumsy. Unconsciously, this willing mind, anxious to learn, simply says, "Okay. If you say so." And it becomes true for her. Another gymnast or fighter pilot or athlete fails to develop. What did we learn as a child, and what do we teach others?

A young boy in a preschool nursery is excited about the opportunity to use crayons to draw the American flag. The teacher passes out the paper and crayons and directs the children's attention to the flag at the front of the room. Like all the other children in the class, the boy begins to draw, creating a beautiful picture, in reds and blues and whites, of the nation's flag. The teacher circulates through the class nodding approval or disapproval of each child's efforts, and as the work is checked off, the children are sent out to play. When she stops at the boy's place, she says, as she has said to others, "This is not accurate. Do it again," and moves on. Not at all disgruntled, the boy draws another flag and smiles because it is very well done, all within the lines, and because he has finished ahead of the other six children who are still working on a second flag. Again the teacher circulates, until there are two remaining to draw the flag a third time. The boy is now perturbed but dutifully studies the flag at the front of the room and tries a third time. After the two are finished, the other child is sent to play, and the boy continues for a fourth try. Finally, after the fourth try, the teacher sputters in exasperation, "Well, it's still not right, Joey, but go on out to play. We can't all be artistic, you know." Thus, the child learns that he is not artistic and that when he doesn't do something right, he is punished and thought of as a bad person.

There are those whose self-image is strong enough to overcome that kind of unconscious learning. There are those who have received other input that allows them to be tenacious in the pursuit of their goals. Van Gogh was continually told by artists and critics alike that he would never be an artist. Teddy Roosevelt was a sickly, asthmatic youth who nonetheless became a championship boxer and broke horses to the saddle on the way to becoming president of the country. Yet far too many others absorb the unconscious messages presented to them and say in their unconscious minds, "Okay, if you say so."

This process continues through childhood and puberty; we learn that no matter what our parents say, sometimes you have to be ruthless, that some people are just going to be bad no matter what, and that if you want to fit in, you'd better go along with the crowd. Indeed, puberty is one of the most brutal times of a person's life. Your body is changing, your hormonal balance is shifting, and you are expected more and more to prove yourself worthy of membership in the adult society. Think back. It can be said that puberty is a combination of the worst of times and the best of times. Even with all the carefree joy of maturing and dating and discovering first love, there is the trauma of broken promises, shifting affections, or being the last in your crowd to "become a woman" or to "score." Doubts and fears run rampant, and in the process, we learn by action and unintended messages to view ourselves as valuable or worthless, worthy or unworthy, normal or a freak.

Those Things We Learn by Observation and Experience

Imagine that you are about to experience the culminating social event of your high school career. It's time for the senior prom. You have the perfect date, you have the perfect outfit to wear, and everything is going to be perfect. You're looking forward to one last major dance with your friends before you graduate. These are people who matter to you. You've all gone through the trials and tribulations of the high school process and survived. You're excited and you're looking forward to this night and to the graduation to follow in little more than a month. And the day of the prom comes.

Rising early, you rush into the bathroom to brush your teeth and shower, and as you look into the mirror, disaster strikes. For there, exactly in the center of your forehead, perfectly aligned with your perfect nose, is the world's most disgusting, ugly, oversized zit. Your whole world comes crashing down. What to do? You call your mother for help. You call your friends for advice. You try makeup. It doesn't help. You consider working on this blasphemous curse with a needle or other instrument of torture, but you can't be sure it won't make it worse. In desperation, you totally redo your hair to cover the crater as best you can and pray that only you and those selected confidants with whom you have shared your misery will be aware of this horrible disfigurement. And what happens at the prom? You spend the whole night staring at the floor, covering your forehead with one hand as nonchalantly as possible, and pretending that you're not really keeping this perfect date of yours at a distance. And you learn by experience. You learn that when things are perfect and you are about to have a truly wonderful experience, some unforeseen circumstance is going to ruin it. You learn that you have to protect yourself from the scorn of others and that in social situations, particularly the important ones, you are going to be imperfect. And you learn that when you have the perfect person in your life at the perfect time, you are going to be highly imperfect, and that will ruin it.

Admittedly, this is a gross exaggeration. Or is it? What do we learn by reaching our own conclusions? When you enter the business world, you may have all the intentions in the world to be valuable to your company and loyal to your boss and to have integrity in your job and dealings with others. Yet your observations may convince you that this is not an appropriate behavior. You may be taught to cheat on expense accounts and that management expects it and condones it as long as you don't cheat too much. You may observe others lying to customers or making false claims or passing the buck by finding others to blame for their own shortcomings. You may be ostracized for putting out too much work, making others look bad, or being too efficient and therefore a "management lackey." Or you may learn that it's okay to go to any lengths to steal customers from a competitor. What do you do in the face of that kind of experience and observation? After all, you're the newcomer here. You're the one that does not yet know what is expected of you. And if the behavior of the firm is different from the professed principles of the firm, whom do you believe? What do you internalize as the corporate culture?

A firm may publicly profess loyalty to its employees, exhibiting paternalistic nurturing of their careers and lauding the accomplishments of individuals in the firm. Yet if the same company lays off workers without notice, fails to reward excellence except with lip service, or manipulates salaries to reflect the budget rather than accomplishment, employees will receive an entirely different message than the intended one. Whether we are talking about business, politics, personal relationships, international affairs, or religion, it is through the experience and the observation of behavior that you receive your sense of reality, either consciously or unconsciously.

Obviously, all learning is not negative. There are many wonderful concepts and workable solutions to life gained through education, unintended messages, and experience and observation. In truth, it is not the event that creates the paradigm but our interpretation of that event. As with the three women in the scenario presented in the beginning of this chapter, it is the interpretation of the event that creates their reality. This is actually good news. It means that no perceived reality is the real one; it is only an individual reality. That gives us a tremendous freedom. As the saying goes, we cannot change reality, only our perception of reality. Fortunately, that is enough to change the entire content of our lives.

The objective reality of concrete objects is not the one that we experience. Our reality is subjective. All information that we receive and upon which we base our idea of reality is brought into our brains through our physical senses. What happens to it once it is there is dependent on our personal perspective. We take that information and filter it through our own fears, beliefs, attitudes, and stored experiences until it is distorted into some pseudoreality that fits into our own world view, our own paradigm. For us as individuals, *no thing is real.* The one and only reality that we can know is our own consciousness about events. Literally, for each individual, consciousness is the one and only reality with which we can deal. And that one fact has a tremendous effect on our behavior, our morality, and our ethics. As we will see in the next section, however, it does not change our values, only the way in which they are expressed.

COUNTERPOINT AND APPLICATION

For our purposes, there is a physical world in which we live and with which we have to deal. It has characteristics, boundaries, and limitations. It has repetitively observable behavior. It is reality. Yet quantum physicists tell us that the only reality that actually exists is in the interactions among elements. There is no physical reality except for those interactions, and it is only through exchange of information that we even know of the existence of anything. Is this just rhetoric designed to confuse us, or do they know something we don't?

Actually, there are two issues here. One is whether the physical world has existence, and the other is whether we experience that true reality. In both cases, the answer is a matter of how we look at it. In other words, this is not a cut-and-dried issue. To the Greeks, everything consisted of combinations of earth, air, fire, and water. To them, that was scientific truth, and they interacted with their world on that basis. To Newtonians, the laws of the physical world and celestial mechanics were provable, demonstrable, and determined. A set of formulae could define exactly what the behavior of the planets in orbit would be, as well as such things as the speed at which an object drops and its velocity at any given point in time. When Einstein came along, he proved that Newton was only dealing with approximations. By the time of Niels Bohr and Oppenheimer, we were splitting atoms and releasing incredible amounts of energy, something the Greeks would have said was impossible. In fact, the very word *atom* means indivisible. It couldn't be split! Who was right?

The answer is, of course, all of them and none of them. They viewed the world differently and, therefore, had a *belief* as to how the world was made, rather than a firm understanding of it. The same is true of us today.

According to the information of the text, if this is true, we should be able to change our world by merely changing our beliefs about it. This should not be a surprise to anyone who thinks about it. Have you never had an opinion about someone and then later found out that the truth was different from that opinion? Have you never disliked someone you did not know well only to find that you liked the person once you got to know him or her?

The same concept is easily applied to technology. The reason we create the technologies we do is that we have a view of the world that allows for them. We could as easily create different technologies with a different point of view. Our paradigms limit our understanding of available solutions by restricting our sense of what is possible. Perhaps there is value in considering some of the "intuitive" ideas that we come up with that do not make logical sense. Perhaps (just perhaps) if we were to pursue them beyond immediate dismissal, we would find they have value we had not understood. To apply the principle to technological creation and use, we need to shift our paradigm and look at things in a different way. The results could be quite astounding, just by redefining the problem or looking for another view of results.

A prime example of this is Post-it notes. Post-it notes are the result of 3M engineers developing an adhesive that was too weak for normal applications. Someone finally decided to quit looking at what was wrong with the product and consider how to use what was there, that is, a glue that did not hold well. The result was a very successful product.

EXERCISES

1. New information creates a new sense of reality for us. In truth, we continually refine and alter our understanding of reality by virtue of our ongoing experiences. Think of the times in your life that you were shocked to find out that something was not as it seemed. How did you feel when you discovered the "truth" about Santa Claus? (He's quite real, by the way. You did know that, didn't you?) What about the first

time you discovered that your parents were not infallible? How about the first time you had a romantic encounter that turned out badly because the other person lied to you? Pick four or five such events from your life and write a paragraph about each of them, recounting what you believed previous to the change in paradigm, what happened to change it, and how you feel about it now.

2. Make a list of at least ten capabilities that you feel you do not have. List things like "I can't sing" or "I can't dance" or perhaps "I just can't get math." Now that you've got the list, look and see how your paradigm is limiting you. Next, pick the one that really bothers you the most and try it. Actually take the time to try it. If you think you are a rotten artist, get some paints or pencils and do it anyway. If you can't swim, go learn. That is a difficult enough task in itself; it involves going against your own sense of reality and deciding to do something that you "know" is impossible. Yet if you give yourself the freedom to try, you will probably find that "Yes, you can!" Now to make it even more frightening, show the results of your efforts to friends. Finally, take some time and think about when and where you first learned that these were impossible things for you. Ask yourself, "Is it true, false, or simply that I don't know?" Now go try it again.
3. If you think back, even superficially, you can probably find instances in your life that were embarrassing, demeaning, or in other ways painful. It is in such experiences that much of our unconscious perspective on the world was created. Consider several such events in your own life. Start with moments of embarrassment and consider the ways in which these events have influenced your behavior and beliefs since then. Now try to view them simply as events without any social consequences beyond what you attached to them. How would your life have been different if you had not decided to feel embarrassed?

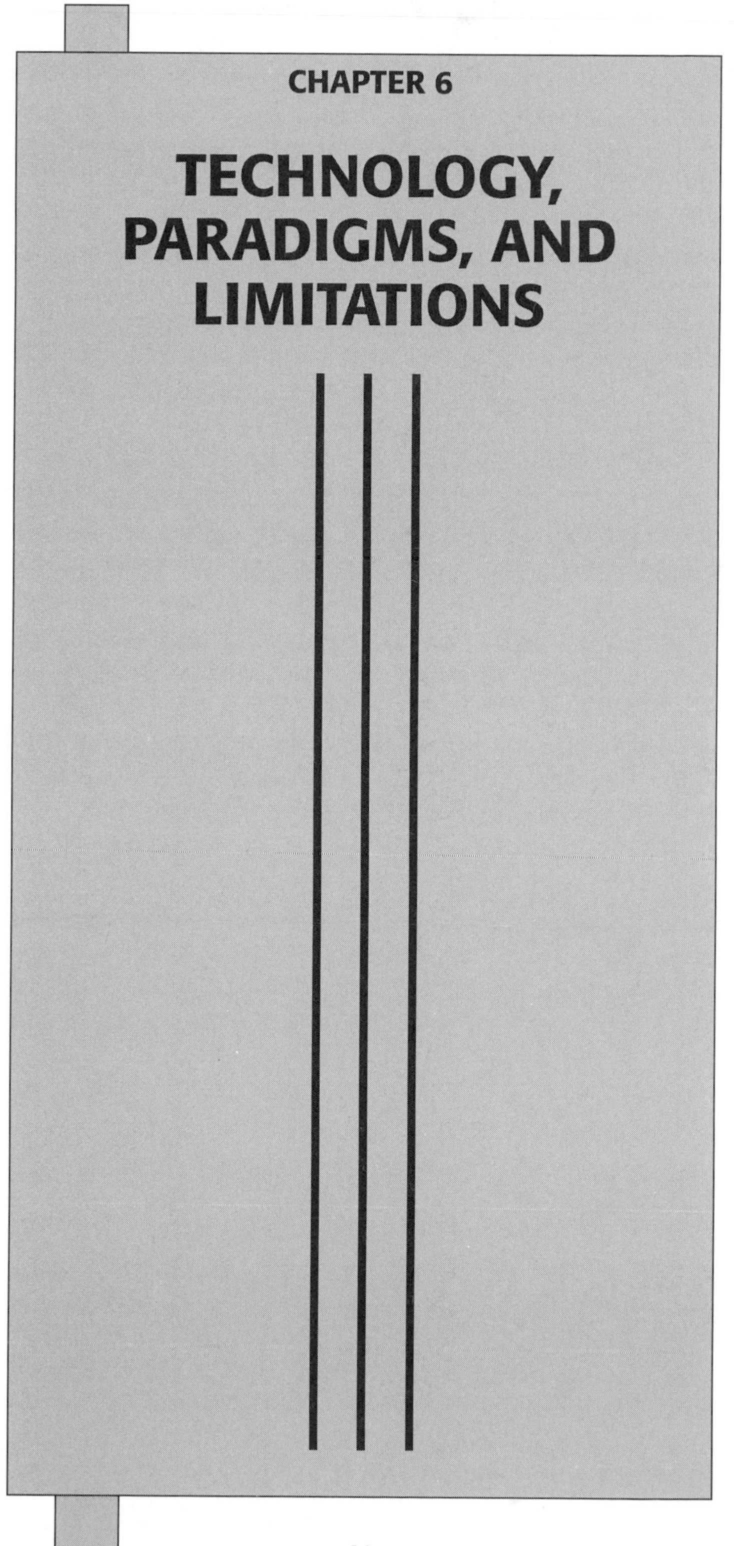

CHAPTER 6

TECHNOLOGY, PARADIGMS, AND LIMITATIONS

PARADIGMATIC SHIFT

In technology as in the rest of life, our paradigms limit our response to our environment and to our understanding of that environment. If ethics is a matter of determining what works and if what we think works is based on our paradigms, then there could be completely different ethical reactions to given situations, depending on the individual's understanding of the content of that situation. The same is true in terms of technology and what is ethical within the technological realm. Consider all of the controversies of the twentieth century brought about by technological development. Many of them are considered something other than a technological issue, yet they still stem from technology, its development, its implementation, and its application. Unionism is considered a social and economic issue, with moral content and a great deal of disagreement in the controversy as to who is "right." Yet isn't unionism in the United States a technological issue? Until there was technology, there was no unionist issue. In the middle ages, craft unions arose within the middle class with the purpose of limiting the number of people who could carry out a given trade to ensure quality and expertise. These artisans, free men capable of producing a valuable commodity, were skilled in the use of tools and technology. As industrialization took place and the separation of the artisan and his tools arose, accompanied by mass production and the reduction of the need for skill in performing a variety of tasks, the unionist movement changed, and the issue became a matter of the treatment of relatively unskilled factory workers rather than skilled artisans working for their own benefit. This was certainly an appropriate shift, considering the monopsonistic tendencies of large industries dominated by a small number of very large firms. Again, it was technology that created the issue, and without the movement to a factory system with mass production, assembly lines, and identical parts (thanks to Eli Whitney), there never would have been a need for a union movement or the moral issues, pro and con, that resulted.

The Protestant Reformation is another excellent example of this change in ethical understanding as a result of a change in paradigm through technology. Martin Luther is considered the father of the Protestant Reformation, yet if it were not for the printing press spreading the word of his actions and protest throughout medieval Europe, the movement may never have gotten off the ground. Why were people ready for such a change? Why were there influential people among the population willing to openly argue with the preeminence of the Church in determining doctrine? It was in no small part a result of the increased availability of books and ability to read among the population as a result of the printing press. Marshall McLuhan has argued that the printing press represents one of the prime movers of the rational philosophies of the Age of Reason, and he is certainly not alone in this assumption.

ETHICAL ISSUES AND PARADIGMATIC SHIFT

In truth, every new technology results in a new paradigm or, more normally, a paradigm shift, and the more extensive the application of the new technology, the more extreme this shift is likely to be. When new possibilities arise in terms of what we can do technologically, it becomes necessary to make judgments about whether those technologies are desirable and appropriate. It has been said that every new law creates a new class of criminals. In a similar vein, every new technology creates a new class of heretics and "sinners." The new technology will represent a divergence from the traditional way of operating in the culture, a way of operation that has been tried and trusted and is known to work for the population. With new possibilities come new questions naturally. Suddenly, issues that never existed before are facing the population. Suddenly, new questions of purpose and the proper use of the new technology present themselves. And as the technology is absorbed into the culture, these questions become more noticeable and more significant.

Beginning in the last quarter of the twentieth century, the population of the United States has had to consider with increasing frequency the question of euthanasia, or mercy killing. Along with this question has arisen the one of ethical suicide. In both cases, there were clearly discernable rules and regulations concerning these topics. It was generally accepted that suicide was immoral and unethical in this culture. Likewise, the concept of euthanasia was seen as heinous and tantamount to murder. There was no question about it. A doctor would not publicly consider the practice of mercy killing. When it did occur, it did so secretly and quietly, and only under extreme circumstances, as in the case of a newborn baby with such extreme birth defects that life was going to be very short and very painful.

What created the current upheaval regarding these issues is a matter of technology. With increased technological sophistication in the health field came the ability to keep someone alive for extended periods of time. Longer life spans led to new health issues, such as Alzheimer's disease and some cancers that would normally not occur to any level of significance in the population. Formerly a rare disease, Alzheimer's is a condition that usually does not affect a person until he or she has reached an advanced age. It has not been selected for elimination as a genetic disease because it does not even appear in most cases until long after the age of procreation. Diseases with a genetic tendency are passed on generation after generation; if the population lives long enough and has a genetic predisposition toward a disease, it occurs at increasing rates. Technology is the reason we are living longer, along with better nutrition. This means that the incidence of Alzheimer's is a technological issue. Thus, it is technology that has created the need to deal with such problems.

Any technology that seriously changes the way we operate in society will create ethical issues, most directly, of course, if the changes in our operation are in direct opposition to our traditional belief systems. Thus, for example, when it becomes possible to both run a household and have a career, the role of women in the society is rethought. When it becomes possible to be fully productive without ever leaving home through telecommuting, we are forced to rethink our concept of work itself as well as what constitutes a day's work for a day's pay. This is an ongoing process. It cannot be avoided as long as we continue to make progress, and judging from the history of the species, that is how we will continue in the future. Thus technology begets ethical issues as a natural result of the changes that technology foments in society.

COUNTERPOINT AND APPLICATION

It can be argued that all technology is the result of paradigmatic shift. Without new approaches, there would be no new devices. The problem arises not so much with the small paradigmatic shifts needed to modify a technological device or application thereof but with the accompanying paradigmatic shifts and larger technological issues. Every time a problem arises with a technology that was not known before, we have technically had a shift in paradigm. To solve those problems, further paradigm shifts are required. The trick is to shift at the time of creation, that is, to develop a facility for and habit of examining each new development at its inception in search of new ways to view the world that include the new technology. This sometimes happens by analogy and sometimes by examination of data. But regardless of how it happens, technological change will always create new worlds for us to live in.

I leave it to you to think about how true this is. Start with the effects of the cell phone or the computer on your view of life. If you are old enough, consider the differences in how you view your life as television has developed and changed through the years. Can you remember a life without computers, without television? Now apply this experience to the burgeoning technologies of today, such as cloning or nano-technology.

EXERCISES

1. Think back to your childhood and the beliefs that formed your view of the world. How have they changed? Specifically, explain how your view of the following ideas has changed since you were eight years old:

a. Parents

b. School

c. The United States

d. God

e. Work

2. Now reexamine your responses to the list of concepts in exercise 1. What has happened to change your paradigm regarding these matters?

3. If we assume that myths are designed to shape the paradigm of children, what type of paradigm was to be conveyed in each of the following? What is the difference between the first two items and the last?

a. *Aladdin and the Magic Lamp*

b. *The Three Little Pigs*

c. *Casey Jones*

4. Find a fellow worker at your place of occupation who is twenty years older or younger than you. Ask the person how he or she felt about going to work and being in the business world when he or she first started out. How does this differ from your own conceptions of work when you began?

5. Consider the fact that paradigms change in many large and small ways on an ongoing basis. How have your views of the world changed since you began this book or class?

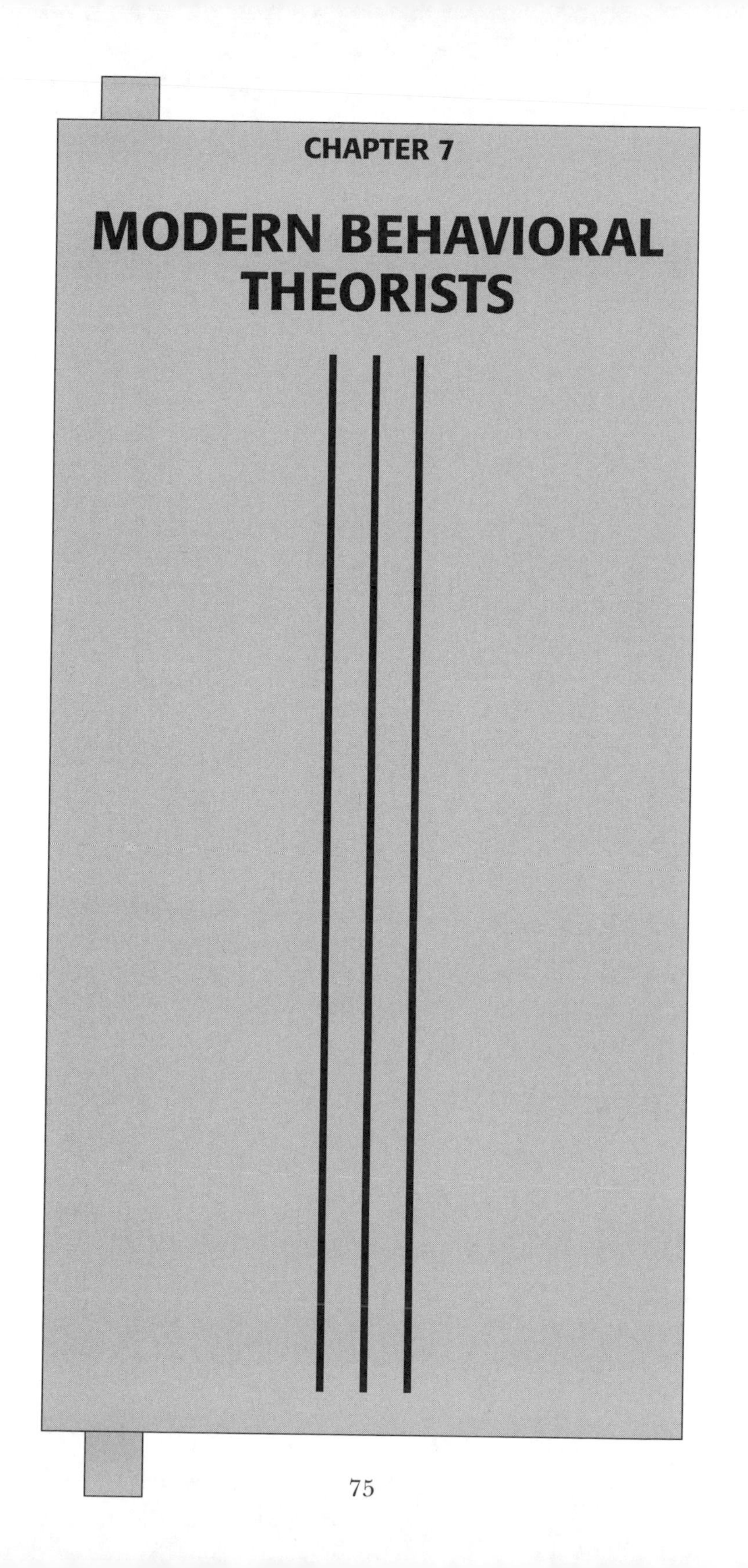
CHAPTER 7
MODERN BEHAVIORAL THEORISTS

INTRODUCTION

In this chapter, we will concentrate on a number of modern theories, many of which are behaviorist in nature. That is, we will be looking at the works of psychologists and ethicists who view ethics from the perspective of an individual's behavior in the world rather than from the perspective of what they "should" be doing. The purpose in this is to determine the reasons for a given mode of behavior rather than to pass judgment on the value, or "rightness" vs. "wrongness," of that behavior. There is an inherent hierarchy of ethical behavior that will be seen in all of these works, however, with an ascending scale of quality of ethical behavior that insinuates what type of behavior is preferred, and this implies judgment. With this in mind, the reader should be aware that however much the tendency toward ethical judgment one finds, the main thrust of the arguments is one of *how* people do perform in the world rather than how they *should*. As the material is studied, this point should become clearer. Suffice it to say that the key to this chapter is the concentration on what motivates a person to behave in a certain way.

The specific theorists that we will look at include Abraham Maslow, John Dewey, Jean Piaget, Lawrence Kohlberg, and Gordon F. Shea. Also included in this chapter is a general discussion of dynamic systems theory and how it also applies to ethical behavior.

THEORIES OF ABRAHAM MASLOW

Most people who have been exposed to the principles of psychology are at least superficially familiar with the work of Abraham Maslow (Maslow, 1968). Maslow's work was not a search for the psychological roots of ethics but rather a search for what makes a person psychologically balanced and successful. He noted that the psychologists who were developing behavioral theories were all concentrating on what makes a person abnormal and apt to exhibit deviant behavior. That's all well and good, but Maslow noted that what is actually normal had never been truly defined. He set out to determine what standard should be used in measuring behavior.

His method was rather simple in concept. He studied people who were generally happy and successful *on their own terms*, whether they were wealthy or humble, well educated or not well educated. He was searching for characteristics that could be applied to success in society. In doing so, he developed what was to become the hierarchy of needs. This theory purports that an individual's behavior is dependent on the satisfaction of certain key categories of needs, which include, in ascending order, physiological needs, security needs, belonging needs, self-esteem needs, and self-actualization needs. For a person to be totally satisfied and completely content, all of these needs should be fulfilled.

According to Maslow's theory, the more basic the need, the more primary the pull for satisfaction. That is, as a lower level need becomes threatened, the move to satisfy the higher level needs is abandoned in favor of the more basic need. A quick look at your own experience will demonstrate the truth of this. If *physiological needs* are threatened (not enough food, water, warmth, and so forth), then *safety* becomes secondary until a sufficient amount of the physical components has been secured to ensure survival. It is of little use to be safe while one starves to death. Similarly, *being a member of a group* is primary to human survival. If one is busily working on *self-esteem* by pursuing status but finds that it begins to threaten membership in one's chosen social group, the tendency will be to ensure that group membership rather than risk losing it just for the sake of self-esteem issues.

The highest level of need fulfillment in the Maslowian scheme, and the most likely to be abandoned if a lower level need is threatened, is *self-actualization*, which is simply a matter of self-expression, of being truly who and what you are. The arts, creative processes, and attempting to develop higher consciousness are examples of attempts to satisfy the need to self-actualize, but if a lower level need is threatened, it will generally be abandoned. As an example, if one is involved in a "spiritual trip" to become enlightened, the behavior may be met with derision by one's friends and family. There will be a great deal of pressure to abandon the project and return to something that the group finds more "normal" and, therefore, less threatening. Under normal circumstances, an individual will simply revert to fulfilling the lower level (and therefore more powerful) need.

Yet Maslow contends that those individuals studied who were truly content and successful on their own terms did not follow this pattern. For those displaying success behavior, the tendency was to continue with their original intent, even in the face of severe need deprivation, whether a matter of peer pressure and derision or actual physical deprivation. There was a refusal in these individuals to succumb to lower level needs. In such cases, the results were fairly predictable: Either the pull of lower level needs eventually became so powerful that the reversion took place or the individual simply continued with his or her self-actualizing behavior and was perfectly willing to forego the satisfaction of those lower level needs. Additionally, Maslow found that those who persisted in pursuing self-actualization were able to secure satisfaction of the other needs in the hierarchy anyway! How can this be?

Consider the case of a theatrical dancer. This profession is highly creative and involves incredibly stiff competition. Yet there are those who are so filled with the joy of expression offered by the medium that they are willing to make great sacrifices in order to achieve their goals. They are willing to take on menial jobs, effectively degrade themselves in "cattle call" auditions, and work long and arduously in an atmosphere of uncertainty. Yet if you ask them why they put themselves through all that, they will reply that it is simply the best job anyone could ever have. They are passionate about their work. They thrive on the energy they gain in the process and in the feeling of satisfaction that they receive from simply doing a job well. They are, in fact, totally successful on their own terms.

Obviously, there are other reasons for becoming a theatrical dancer. There are the lures of wealth, fame, and recognition. But most of those who succeed and are happy with their choice are not motivated so much by those factors as they are for the pure love of what they do. And those who do receive the wealth, fame, and recognition can be miserable with it if they end up asking
"Is that all there is?"

Maslow breaks his hierarchy into three distinct categories: *physical needs* (which include both physiological needs and safety needs), *social needs* (which include belonging and that part of self-esteem that depends on acceptance by others), and *higher level needs* (which include true self-esteem, which comes from within oneself, and self-actualization). He sees individuals as being trapped by their needs and desire to fulfill them, except when they are about the process of self-actualizing. Only in this stage are they truly free to simply be who and what they really are, without any desire to protect their other needs. We will return to this concept later in the text to see how it fits into the scheme of behavioral ethics. To better understand the concept and, thus, facilitate that process, it is necessary to mention the peculiar dual nature of self-esteem.

Self-esteem can be divided into two types: *external self-esteem* and *internal self-esteem.* In external self-esteem, the feelings of well-being stem from the opinions of others. In internal self-esteem, the feelings of self-worth come from within the individual. In external self-esteem, individuals are concerned with the opinions of others, about how these others view them rather than how the persons feel about themselves. It is an attempt not only to belong to a group but also to stand out in the group. In the case of the theatrical dancer, the lure of fame and adoration by others is in the category of external self-esteem.

With internal self-esteem, the dynamic is quite different. This is where the passion lies, where individuals begin to feel good about themselves because of the type of work they are doing and the fact that they are doing it well. Even if they are not the best of professional dancers, they still view themselves as growing and developing, as improving and sharpening their skills, and thus doing their best. Internal self-esteem thrives on the quest for one's best, whatever that may be, and is satisfied by an honest effort to achieve that goal. In order to have internal self-esteem, individuals find themselves in a condition of honesty, integrity, and a willingness to take personal responsibility for their own actions. There is no consideration of others' opinions of the individual in the determination of actions here. The individual takes full responsibility for personal actions, a key ingredient in being truly ethical, as we shall see. Such persons are honest with themselves and others as to what they want, what their motives are, and where they are trying to go. Such persons have integrity about what they are doing, that is, they say what they mean and mean what they say. With such a combination of behavior, the person begins to trust his or her own judgment, thus eliminating the need for external approval. This second approach to creating self-esteem is representative of higher need fulfillment. It leads to behavior that is solid, unambiguous, and straightforward. Interestingly enough, what tends to follow is a feeling of admiration and acceptance in others, which supports and fulfills the belonging needs of the individual as well.

THEORIES OF JOHN DEWEY

John Dewey is considered to be a *pragmatist* in that his approach to ethics centers on doing what works to solve the problem at hand (Shea, 1988). He notes that since you seldom know at the outset exactly what will solve the problem at hand, it is a learning process. He feels therefore that ethical decisions must be based on experience to be applied before the fact. He notes that what to do is not a known quantity but rather an action to be taken and then reviewed for effectiveness. To Dewey, the process of becoming ethical is experiential and experimental, and he believes that in order to behave ethically, you must act, experience the process, observe the results of the action, and then pragmatically determine what is and is not ethical. From this practical (pragmatic) point of view, you must scientifically determine ethical conduct through experimentation and observation.

Notice how much of Dewey's approach aligns itself with twentieth-century pragmatism, further supporting the notion that ethical approaches are a reflection of the paradigmatic understanding of the practitioner as to the nature of the world. Just as a medieval theorist would base ethical decisions on the "Word of God," Dewey turns to the scientific method or secular twentieth-century America for a structure in which to make ethical decisions. This in no way negates the usefulness of the approach; it simply helps understand his motivations in his choice *of* approach. It also serves to explain, at least in part, the attractiveness of this approach today to people involved in technology and science.

This behaviorally based approach details Dewey's own understanding of people's behavior. He divides ethical behavior into three categories: *preconventional ethics, conventional ethics,* and *postconventional ethics.* According to Dewey, each of these types of morality centers on specific types of need fulfillment and can be compared (as it will be later) with Maslow's hierarchy of needs.

In Dewey's view, the preconventional approach to morality centers on what Freud would characterize as "idic" behavior, that is, seeking pleasure and avoiding pain, and in what Maslow would call survival needs. This type of morality is based on whether or not the behavior satisfies immediate needs. The desire to be ethical in this case, that is, the tendency of the action to satisfy needs and therefore be carried out, is predicated on its ability to reap immediate rewards and lasts only as long as the need is unsatisfied. When it is completely satisfied, there is no longer any motivation for the behavior, and the behavior of the individual shifts to some other pleasure-pain principled issue. In other words, with preconventional morality, behavior lasts only as long as it is useful to satisfy needs.

Preconventionally motivated individuals are likely to do or not do something based on whether they might get caught or punished. With children, for example, the perceived value of cleaning up their rooms or doing a chore is based on fear of punishment if they fail to do so. When they are no longer afraid of punishment for not carrying out some chore, they will lose their motivation to do it and would find this a perfectly ethical decision. Likewise, they may decide to help a sibling or parent because of expected rewards but be unwilling to do so if no reward (need fulfillment) is forthcoming as a result of their behavior. Again, they would see this as ethical and proper if they are preconventionally motivated. This argument is by no means limited to children. Adults may exhibit this approach as well. Such is the case of a worker who considers it his duty to only do those things in the office that are specifically in his job description or that will lead to immediate reward. If fellow workers are in need of assistance, the only motivation for the preconventionally oriented worker in deciding whether or not to lend a hand is dependent on how he is directly affected by the situation, either positively or negatively. "It's not in my job description" or "That's not my department" are typical responses to cries for help to preconventional workers. Theirs is a strictly *utilitarian* approach to decisions. "What's in it for me?" or "What's it going to cost me?" are the questions such workers ask themselves.

In the conventional stage of ethical development, morality is determined on a different basis. Dewey sees people who are conventionally ethical as motivated by a desire to be a member of and participate in the structure of their social group. Their decisions stem from their perceived place and role in the group and are dependent on the morality/ethics of that group. Hence the emphasis here is on what Maslow would classify as social needs, including belonging and externally secured self-esteem based on the attitudes of others toward the individual. Conventional morality and ethical behavior are based on the expectations of the group and the rules and regulations under which it operates. It is that simple. In this mode of operation, the individual will act and continue to act in a given way as long as the group indicates that that behavior is acceptable. The problem with this arises as the group changes and, thus, the understanding of what is and is not acceptable changes, though the official rules of the group may not change to reflect this shift. This is where, for instance, generally accepted behavior is found to be in opposition to the law (official group norms) or formally accepted modes of behavior. An example of this would be individuals who, by virtue of their membership in a given group, may choose to speed on the expressway when the flow of traffic is speeding, though the law says it is not ethical to do so. Note also that as the affiliations within groups change for a given individual, that individual's behavior may also change in response to those groups' concepts of ethics.

For this last point, consider the indoctrination process undergone by recruits into the armed services. It is designed to realign the ethics of individuals entering a new and different group. The process of basic training, along with its goals of physical conditioning and teaching the military way of doing things, is to take a widely disparate group of individuals, most of whom are deeply imbued with the concept of killing as an ethical taboo, and convince them that it is okay as long as the reasons are in line with those of the group. People are taught to not think for themselves but rather follow the orders of their superiors, to carry out irrational acts (such as charging machine guns or shooting other people if necessary), and to fit into a new culture centered on the use of this potential force on command. The recruit's personal point of view becomes that of the organization (group) with which he or she has become affiliated.

A more mundane example can be found in a business situation in which a person accepts the "corporate culture" with all its inconsistencies of conduct (illegal use of copy machines, padding of expense accounts, use of company time and equipment for personal use) based on the belief that "it's all right because everybody does it." Or an example can be found in a technological environment in which potentially negative results may come out of research and development because of beliefs such as "science doesn't do harm, but people do harm with how they use the technology we develop" or "I know it's dangerous to the health of the public, but how it's used is not my worry; I'm paid to come up with the ideas."

These are conventional views. If the group approves it, either officially through rules, regulations, and laws, or tacitly by unspoken mutual agreement (looking the other way), then it's ethical and okay. Again, the emphasis here is on following the lead of the group in determining what is and is not ethical.

In the case of postconventional morality, Dewey sees people as making ethical decisions based on a personal sense of morality, incorporating not only their intellectual, thinking processes in the decision but also a more personal, feeling process as well. Ethical decisions made in this mode are closely connected with what Maslow would call personal development needs, that is, internal self-esteem and self-actualization.

Postconventionally centered ethics is decidedly different from the other two approaches. For the first time, in the postconventional approach, the individuals take responsibility for their own decisions and do not base them on merely the satisfaction of momentary desires or the concepts of others. The decisions are made on the basis of personal views about what is and is not ethical. It is an internal process, by which we make our own decisions and do it autonomously based on an understanding of what works. Thus a person operating postconventionally, though given the chance to cheat on a test and operating in a culture in which cheating is generally accepted, may still not choose to do it. And the decision is not based on fear of being caught (preconventional motivation) but on the basis of it not contributing to his or her learning. Or perhaps it is on the basis that whether accepted or not, it is a lie and the person believes in telling the truth. Consider also the worker in a production environment in which rapid production without regard for quality is acceptable within bounds, who still insists on doing quality work, even at the expense of time. The same is true of a tenured college professor who knows full well that he won't be fired for slacking off in the delivery of material to students and still, on the basis of personal principles, works hard for the students. In each of these cases, the ethical decision is a personal one, exclusive of others' opinions and, in some cases, painful in the short run.

Dewey feels that people in all three groups may have the same behavior patterns, but their reasons for the decision to behave in a certain way are based on entirely different motivations. It is, then, the motivation that is the key to determining the level of ethical development on which a person is standing. Students may decide not to cheat because they are afraid of getting caught (preconventional) or because the group frowns on cheating (conventional) or because they do not believe it is ethical to cheat in the first place (postconventional). The behavior is identical, but the motivation for that behavior is different in each case. It is *not* in the behavior that the morality of an act lies, but in the motivation for that behavior. Dewey sees people in the preconventional category as morally lacking and limited. Those in the conventional category are seen as having turned over their moral decisions to others and are therefore basically nonethical in that they have no personal basis for their ethical decisions. Those in the postconventional category can be seen as autonomous and willing to be responsible for their actions, thus exhibiting a much higher degree of freedom of action than those in the other two categories. The conclusion arises that many people live within limitations of behavior due to the nature of their morality and that personal, autonomous morality offers the greatest degree of freedom of action.

THEORIES OF JEAN PIAGET AND LAWRENCE KOHLBERG

Lawrence Kohlberg, building on the work of Jean Piaget, developed his own hierarchy of ethical development (Kohlberg, 1981). His model of ethical behavior stems from the work of Jean Piaget, who worked in the field of child development. Strictly speaking, Piaget was operating in an entirely different line of research, studying how children develop over time, and was seeking to use this information to determine appropriate behavior and appropriate learning strategies for children of different ages. He categorized child development into six distinct stages, and Kohlberg, also working in the field of education, extended the theory to encompass the development of a child's ethical behavior as well. From this work, we find another behavioral hierarchy to apply to our understanding of why all people, not just children, make ethical decisions the way they do.

Kohlberg's view is that moral development is a learned phenomenon. At birth, he reasons, all humans are free of an ethical structure, the concept of morals, or, for that matter, honesty. He sees the family as the primary source of values and morals as they develop in the individual. Patterns of moral behavior, he reasons, change as a person's intelligence and ability to interact successfully with others increase. He coupled this with Piaget's *morality of cooperation* theory to develop his own theories.

Piaget's concepts of ethical development among children represents what is essentially a learning process that allows the child to shift from what he calls *moral realism,* a condition in which the young child views rules as immutable and set. When Piaget looked at the moral development of children, he found them to be "imprinted" through many responses by their parents, and they automatically reacted in ways that were preprogrammed responses to behavior. The children had taken on these responses as the indication of what is and is not moral. Any action that is defiant of these rules is punishable; there is no distinction between degree of offense or reason for the offense.

Slowly the child develops a view of morality that Piaget characterizes as a *morality of cooperation,* in which the child begins to understand that rules are created by individuals and groups of individuals for particular reasons and that these rules can be changed by individuals and groups as well. In this stage, there is a better understanding in the more mature child of the relationship between motivation and action, which leads to some interesting dilemmas in the child's mind as the child attempts to sort out when to behave in a given way and when not to. Punishment is also seen as variable, again dependent on not only the degree of damage done by breaking the rules but also the motivation for doing it.

Continuing with this work, Kohlberg investigated the nature of morality of people and developed his own hierarchy of six stages of moral development. Kohlberg believed that people do not have any morals or ethics at birth but rather that they develop morals through these stages as they mature. It is understood that they may not go through all of them and that they may, in any particular moment in time, be in any of the levels of the hierarchy.

Kohlberg divided his hierarchy into three main categories: preconventional, conventional, and postconventional. Then he divided these three categories into two groups. The resulting six levels of moral development are as follows:

1. Punishment-obedience orientation
2. Personal reward orientation
3. Group norm orientation
4. Law and order orientation
5. Principled morality and social contract orientation
6. Universal ethical principle orientation

In exploring each of these orientations in detail, it is good to keep in mind that there are those who disagree with Kohlberg. This is particularly true in terms of whether or not the orientations are experienced in discrete steps or as a continuum and whether the reasoning (orientation) individuals use to make moral decisions matches their moral behavior. With these criticisms in mind, the six stages presented by Kohlberg are discussed further.

1. *Punishment-obedience orientation.* In this stage, ethical actions are seen as based on the avoidance of punishment and pain. In the punishment-obedience mode, the person making the decision sees any act that can reasonably be expected to be free of punishment as moral and any act that leads to punishment as immoral. Thus, as with a small child in Piaget's moral realism stage, the individual simply follows orders and obeys the rules to avoid negative consequences.

2. *Personal reward orientation.* The second stage in Kohlberg's scheme revolves around the reception of pleasure (reward) rather than the avoidance of pain. If an act results in benefits to the individual, it is a good act, and if it does not, it is not a good act. This is still part of the "idic" pain-and-pleasure principle of Freud's psychology, but its level is a bit higher than the purely fear-oriented philosophy seen in stage one. Behavior fitting this orientation can be seen in the self-interest of capitalistic market structures, that is, economically rational people always try to maximize profits or, in a general form, maximize pleasure and minimize pain. It is a rather sad commentary on our world that most of the business sector seems to work within this market morality approach, though, as we will see later, there is a serious price to be paid for such an approach.

3. *Group norm orientation.* In the third stage, the determination of what is and is not considered ethical shifts from one of punishment and reward to one of social awareness, in that the determination of morality becomes dependent on the attitudes and mores of the social group to which an individual belongs. The individual begins to look at what the social structure as a whole deems acceptable. That which is considered ethical becomes based on the group decision about what is ethical. Note that this is an informal and nonofficial form of moral code, in which individuals observe what is actually tolerated by the society or considered acceptable by the group with which they are affiliated. With membership in multiple groups, behavior shifts with the norms. What is socially acceptable in a work environment or family environment may not be (indeed, probably is not) acceptable in casual social situations. Morality thus becomes multifaceted and changeable as the group changes.

4. *Law and order orientation.* At this level, the basis for ethical reasoning is what the stated rules are, and behavior is dependent on those formalized rules. Effectively, it is a step from conforming to group norms in that it takes as the measuring device the official, social statements of what is acceptable or not acceptable rather than the behavior patterns of individuals and groups. Groups within a society may or may not behave morally in this level of understanding, but the rules of the society are certainly a statement of what is ethical and what is not, according to acknowledged group beliefs. Law and order orientation eliminates a number of behaviors that might be acceptable "in the vernacular," by stating the official line. It may also limit a person's ability to respond effectively in a situation that is not clearly defined by law or that may have extenuating circumstances. As with the case of depending on group norms, this stage creates inevitable moral dilemmas when circumstances do not exactly fit the definition under the law. Changing conditions create changing responses, and as a body of codified responses, laws may not be able to appropriately react to a given set of conditions. Yet those who depend on this level of reasoning have at their disposal a method of making moral determinations that is exact and precise: If it's legal, it's moral; if it's not legal, it's not moral. Needless to say, the amount of room for rationalization in this stage is as great as in the previous three.

5. *Principled morality and social contract orientation.* At this fifth level, the individual begins to take personal responsibility for his or her actions. Principled morality is morality based on autonomous decisions and responsible choices. Here, the basis for reasoning and decision making is internal rather than external or circumstantial. Responsibility for one's own personal actions is considered the key to the determination of what is ethical. Until the basis of determination becomes personal responsibility, the reasoning is based on outside forces, whether it be punishment, the level of profit received, or the opinions of other individuals and groups in the society. It should be remembered that for purposes of ethics, the word *responsible* means "the ability to respond." When you are dependent on others for determining your morality, the ability to respond to circumstances and situations is automatically limited. Only through autonomous, principled decision making can a true ability to respond be realized.

In terms of a social contract, there is an implied agreement among the members of society that they will behave in a way that is consistent with creating stability and a reasonable degree of satisfaction for everyone involved in that society. That is, rather than simply doing whatever one wishes for one's own good, there is a need to behave in light of the effects that behavior will have on others involved, since to not do so would result in chaotic circumstances. If everyone ignored the effect of their actions on others, there would be no stability or order in the society at all. Hence, it is imperative that the results of our actions not harm others involved.

Again, the emphasis on reasoning and decision making shifts from what others think or may do as a result of our behavior to a conscious, rational decision at which one personally arrives, regardless of the undue influences of what others think.

6. *Universal ethical principle orientation.* At this final and highest stage of ethical development, Kohlberg asserts that an individual begins to understand and see the universality of ethics, that there is a set of ethical principles that are immutable and universally applicable. At this highest level of ethical development, the individual sees what is ethical as being what is ethical for everyone concerned, and everyone is taken into account in deciding the ethical nature of the decision. At first inspection, this seems to be simply a matter of social contract, yet it is different in that it does not involve compromise as much as mutual benefit. Rather than modifying our behavior in accordance with the wishes of others, we look for a solution that is truly ethical for everyone. No one has to lose anything, but rather everyone benefits. This is not a compromise, in which people negotiate to maximize return and minimize loss. It is a search for a perfect solution; only a solution that yields what everyone wants is seen as the ethical choice. In other words, we are faced with the concept of win-win. If anyone loses, everyone loses. If anyone wins, then by definition, everyone must win. There are no losers in the process.

For Kohlberg, this is as highly developed as a person can be ethically, when the person begins to view his or her decisions in terms of win-win or lose-lose. Yet according to Gordon F. Shea, there is a yet higher level of ethics available to us.

THEORIES OF GORDON F. SHEA

Gordon F. Shea is a management consultant and researcher who supports the ideas of John Dewey and Lawrence Kohlberg, yet he takes their process a step further (Shea, 1988). Shea has no disagreement with the threefold hierarchy of preconventional, conventional, and postconventional ethics. Nor does he disagree with Kohlberg's six levels of ethical development. His unique contribution is the recognition of yet another stage of ethical development, one that surpasses the sixth in degree and understanding. It is this seventh step that we now explore.

Shea refers to the seventh step as *transcendent ethics*. He says that there is indeed a final stage in ethical development that moves beyond not only the Kohlberg levels but also the concept of ethics itself. His theory deals with the twin concepts of transcendence and *integrity*. In his approach, integrity is seen as the integration of thought and feeling to make decisions that are creative, caring, and sensitive. This is a mixture of the intellectual and the intuitive in decision making. He explains that the only way this can be done is to remove one's self from the rationale. The developments of earlier models are essentially a matter of conscious, rational choice. Shea says that there lies a stage beyond the rational that is not a matter of logic as much as a matter of intuitive sense regarding what is right and wrong. Essentially, he is referring to what in earlier chapters of this book is referred to as our *values*, that innate sense of what is and is not okay. Such a method of ethical behavior would require a wholly balanced and fully integrated personality that could integrate the feeling nature and the thinking nature without conflict. This is what Maslow would call self-actualization, or a person expressing who and what he or she truly is. In other words, we are talking here about perfect, absolute conscious honesty. Hence the applied term, *integrity*.

The word *integrity* comes from the Latin and means "undivided" or "in touch with." Either base meaning fits Shea's interpretation. To be undivided in the ethical sense is to be free of conflict in understanding of what does and does not work. It is an internal integration of the personality as opposed to the *fragmentation* experienced by most people, who live in an internal world in which competing factions are constantly in a battle for expression, supremacy, and control. People of integrity in this sense are without such internal conflict. They really know who they are and what they hold to be true. Their reality, in other words, is cohesive and noncontradictory.

To be in touch with your true Self is equally applicable. When you are really in touch with your Self, you have integrity. There are once again no hidden agendas, no conflicts, no confusion as to the nature of what is and is not true, but a totally personal view of the world that is free of undue influence from other's ideas, concepts, and programs. When you are in touch with your Self (true Self, capitalized to delineate it from the fragmented, dysfunctional self), decisions become easy and unconscious. Shea notes that integrity has long been considered the measure of a truly ethical person and that its definition indicates this. According to Shea, integrity is "an uncompromising adherence to a code of moral or other values. It involves utter sincerity, honesty and candor and the avoidance of deception, expediency, artificiality or shallowness of any kind" (Shea, 1988, p. 88).

The implication here is that the individual with integrity has been completely freed from the chains of Maslow's hierarchy of needs and is just being totally Self. Such a person would naturally behave in an ethical, honest manner no matter what, since he or she would have neither fear nor an inordinate desire to satisfy lower level needs. Only when free of these fears is one indeed in a position to make totally ethical decisions. Additionally, those decisions come automatically, without any need for consideration, contemplation, or rationalization.

This all sounds like a very tall order. It may, in fact, appear to be pretty much impossible. Yet we all have some level of integrity, and we all have moments when we find ourselves behaving ethically without giving thought to the conscious choice to do so. It "just comes naturally," so to speak. At these moments in time, you are operating honestly from your heart and are in integrity with your true nature. You are self-actualizing.

Consider the times in your life when this has happened. If you look closely, you will find times in your life when you have done the correct thing (what truly works) without having been consciously aware of it. It is simply that no other course of action ever entered your mind. Opportunities arise every day to do unethical acts, for example, not pointing out that you were given too much change in a purchase. Yet if such opportunities are not only not taken but also never even considered, then this is an act of integrity. This is an ethical act. This is definitely doing what works for everyone.

Shea's transcendence literally goes beyond the whole idea of making ethical decisions. If unethical choices never arise in the individual's mind, there is no choice being made. Action is natural. But how can this state come about? How does a person achieve this transcendent state? Basically, the way to approach Shea's seventh level is to simply become aware of the truth that lies within, at the level of core values, and to understand the manner in which the universe functions.

A COMPARISON OF THEORIES

In figure 7-1, we can more clearly see the relationships that exist among the theories of Maslow, Dewey, and Piaget and Kohlberg, and how they each reflect different schema for expressing the same perceived truth. Shea's transcendence, it should be noted, aligns only with Maslow's highest level in the hierarchy, that of self-actualization. Also note that the order of the various levels has been reversed in the Dewey and Piaget-Kohlberg models to more properly align with Maslow's hierarchy, which is traditionally presented with the highest level first, as in the pentacle of a pyramid.

Kohlberg discusses taking into account how our actions affect others in making ethical decisions. This is entirely bypassed in Shea's seventh level, because we cease considering the effect of our actions

MASLOW'S HEIRARCHY OF NEEDS	PIAGET-KOHLBERG HIER-ARCHY	DEWEY'S CONVENTIONS
Self-Actualization ←	Universal Ethic	Postconventional
Self-Esteem ←	Principled Morality	
Belonging to the Group	Law and Order Conforming to Group Norms	Conventional
Safety ←	Market Morality	Preconventional
Physiological Needs ←	Obeying Orders	

FIGURE 7-1 A Comparison of Maslow's, Piaget-Kohlberg's, and Dewey's Approaches to Explaining Ethical Behavior

on others and begin to realize not only that we affect others but also that we are directly connected to them. We are part of a greater whole, and any action taken toward another human being is an action taken toward ourselves. An understanding of the meaning of this requires that we first discuss the next theory, that of systemics.

COUNTERPOINT AND APPLICATION

The primary criticisms of all of these theories stem from the apparent exceptions or criticisms of methodology in their development. It is not the purpose of this book to discuss or offer apologies for these individual theories. That is not the point. The theories are in the realm of the social sciences, which have less quantitative content and more qualitative content than the so-called hard sciences and, therefore, are open to debate to a high degree. None of them are conclusively proven. If they were, they would be laws rather than theory. Again, this is not the point.

The point is that there is a common observable thread running through all of these theories, a thread that is recognizable and universally experienced by those who have taken the time to look at their own behavior and the behavior of those around them. There are different paradigms in terms of how to determine ethical behavior, and we do observe some as more highly developed than others. The value of the material is that it gives us a map, a guideline for how to improve our ethical standards in a way that works for us. I personally do not know anyone who works on the level of Shea's transcendental ethics all the time. I do know times when I have done so and when most people I know have done so. My suspicion is that most people achieve this level of ethical development if they have gotten beyond level four on the Piaget-Kohlberg hierarchy.

Applying this material is a matter of considering outcomes and working on a mutually advantageous solution to all problems. The solutions are there. Because we do not see them, it does not follow that they do not exist. There is an infinite number of solutions to any problem if we will simply open ourselves to the possibilities and use our intuition.

EXERCISES

1. There is an inherent separation between action and motivation in the behaviorist models. For any given situation, the behavior of an individual in each of the states that a specific

behaviorist is discussing may be totally different or identical to the behavior of an individual in some other place in one of the hierarchies. Hence, two people who choose not to take the opportunity to steal when it presents itself may have totally different motives. In the scenario that follows, determine how a person in each stage of development would behave, according to Dewey, Maslow, and Kohlberg. List the responses of an individual in each stage of moral development.

A man enters a convenience store to purchase a newspaper and a soda and encounters a clerk counting out the daily receipts at the register. The clerk is suddenly called to the back of the store, summoned by a loud crashing sound and calls for help from another employee. He leaves the pile of money, mostly large bills, sitting on the counter and rushes off to help his fellow worker. The customer sees the money, realizes that the clerk has never looked up and cannot identify him, and notices that there are no other customers in the store. He even notices that the security camera has been removed for servicing. What does he do?

2. There are many historical examples of individuals who, by virtue of dwelling in a state of self-actualization, have suffered hardship for "their art." Artists, musicians, composers, sculptors, and writers as well as highly placed business executives and other public figures may fit this category. Yet not all people who are successful are equally motivated to express themselves. Pick three such examples of successful people with whom you are familiar and briefly analyze the results of their efforts. After all the pain and suffering in getting to a position of success, do you perceive them as happy and content (having peace of mind) or as still concerned and striving? What does this say about their motivations for pursuing a given career?
3. Do an honest assessment of your own behavior. You will undoubtedly find you operate in different developmental states at different points in time or regarding different aspects of your life. Returning to the Piaget-Kohlberg model, think about times in which you dwell in each of the states of development. What motivation keeps you there in your decision making? For your own benefit, briefly discuss your findings on paper.

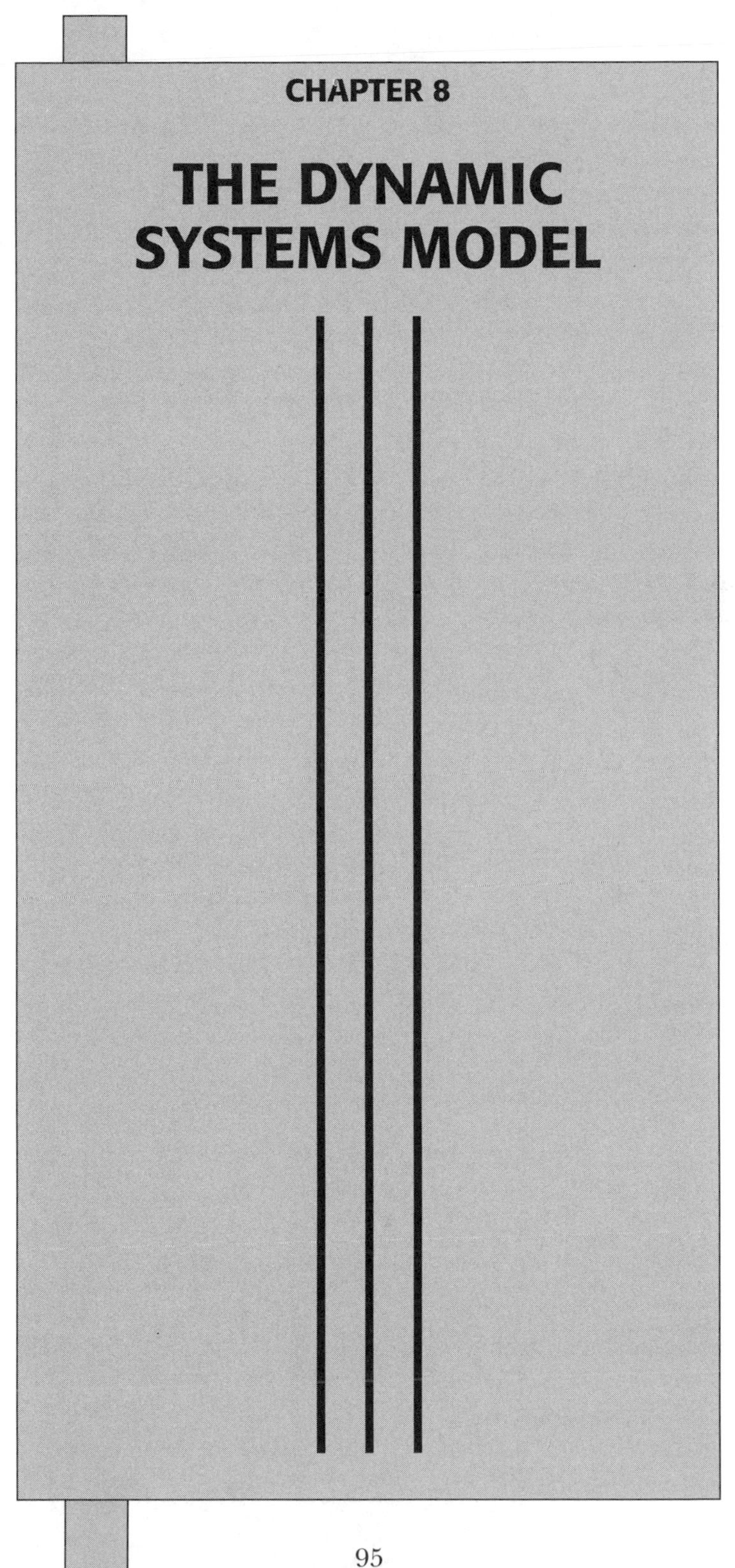

CHAPTER 8

THE DYNAMIC SYSTEMS MODEL

INTRODUCTION

From a technological standpoint, any system of ethics that is effective needs to be both practical, that is, readily implemented, and reflect the reality in which we live. The dynamic systems model allows us to develop an ethical approach that fulfills both of these requirements. First of all, it does reflect the world in which we live. It is a model that describes the composition and behavior of the universe in a manner consistent with experience. Secondly, because the application of the behavioral laws of this model is straightforward, it can be applied to individual decision making and "real world" issues.

The physical universe is an incredibly efficient system. No matter where we look, there are indications of the movement toward order, even in the face of entropy and chaos. To understand how this is true and to develop a method of using the order inherent in the universe, the dynamic systems model has been developed. It describes both the structure and the behavior of the physical universe to a high degree of accuracy. In development since the mid-1950s, the model had undergone a number of changes and, in doing so, has shifted from popularity to disfavor and back to popularity. This is just a matter of development, in which the original flaws in the model were corrected so that now we have a universal model that functions well to describe how things truly work. The model still has critics, but as it applies to our present study, it is eminently qualified as a workable theory.

THE NATURE OF SYSTEMS

There are three primary elements to look at in the dynamic systems model: its definition, its structure, and its behavior. The first two are relatively straightforward and obvious; the third requires a bit more explanation.

Definition of a System

A system is a collection of elements that interact to achieve some goal or create and maintain some state. This is all there is to a system. Yet if we look closely, we can see the seeds of description for any physical object and its behavior wrapped up in this deceptively simple statement.

First of all, we are discussing a collection of elements. Only experience and observation limit what those elements can be. They may be physical objects, groups of objects, or structures of understanding, knowledge, information, and arrangement. An automobile can be an element, as in the transportation system, which involves many automobiles. A committee, a collection of elements, can also be an element of a system, and that particular element involves not only the individuals on the committee but also a host of other inputs, such as data, structure, organization, and so forth. Whatever the nature of the elements, they represent the parts that make up the system.

Secondly, the definition states that the elements interact. Actually, since there is a desired goal or state to be achieved by the system, we can truthfully state that the elements not only interact but also actually cooperate to achieve that goal or state. Yet the word *interaction* is used to avoid confusion, as the cooperation exhibited by the elements of a system may not always be apparent. In a football game, for instance, each team obviously cooperates to achieve its goal, yet the two teams are in competition with each other. Yet if we look at the system represented by the game itself, both teams cooperate by competing to create an exciting contest that follows specific rules and offers the fans an opportunity to enjoy themselves and be entertained. This cooperation between competing teams does not look like a cooperative effort at all, though it is. What happens if they do not cooperate? What happens if one team refuses to show up for the game? What if, upon arrival, they refuse to play by the rules? Would the purpose of the sport be fulfilled? Similar examples can be found in economic theory, business practices, and politics. In fact, nature itself cooperates through an evolutionary process involving competition and survival of the fittest to achieve a state of success for the entire ecosystem. No matter how competitive the system may be, the elements are cooperating for the success of the system itself.

Finally, we have the goal or state to be reached and maintained. Every system has a goal. This goal, though it may be far different for each system, is the very raison d'etre for the system to be there in the first place. There are no systems without a goal, and since the universe is a system made up of systems, there is therefore nothing in existence that is without a goal or purpose. Nothing, no thing, is worthless and without purpose.

Systemic Structure

The structure of all systems is the same in principle. Each system is made up of elements, which are in themselves smaller systems, called subsystems. Think about it. If everything in the universe is systemic and all systems are made of elements, then the elements of which they are made must also be systems. Thus the automobile is part of a system we call the transportation system, along with other elements such as busses, trucks, airplanes, trains and ships, roads and bridges.

Simultaneously, it is a system in its own right, consisting of a braking system, an engine, a transmission, a cooling system, an electrical system, and a wealth of others. Each of those subsystems is in turn made up of subsystems.

This process of nested systems has two key features. First of all, an element of a system, particularly a social system, can be part of more than one systems, as in the case of individuals who are a part (subsystem) of a family, a business through their jobs, a community, a nation, a culture, and a species occupying the planet. The second feature is that each element in the system is both affected by the larger system (the larger system's behavior partially determines the behavior of the subsystemic element) and affects the larger system (the subsystem partially determines the behavior of the larger system through its interaction with other subsystemic elements in that larger system). This will be shown to be a crucial point later in the discussion.

The Three Behavioral Laws of Systems

There are three laws of behavior that control the way systems operate: the laws of synergy, reciprocity, and balance. These laws are universal, immutable, and always function in any system. No matter what the appearance of a system, which is a function of its elements, its interactions, and its goals, the behavioral laws will determine how it operates. Those laws are as follows:

1. *All systems are synergistic.* This first law of systems behavior states that all systems exhibit synergy, that is, they are constructed so that the whole is greater than the sum of the parts. As such, there is some element crucial to the definition and behavior of a system that does not exist unless the separate elements of that system are combined to create the system. Consider the example of a political system. It primarily consists of people and their interactions. Yet people in any group without the interaction inherent in the system of which they are a part are merely individuals. It is only through their functioning as a system that any dynamic systemic behavior takes place. Until they decide to elect officials, implement order and rules of acceptable behavior, and cooperate in the carrying out of mutual goals, they are not a political system at all but rather just a collection of individuals behaving in random ways. It is that dynamic of interaction that creates the political system, not just the people involved.

An office building can be seen in the same light. It consists of steel and concrete and wood and mechanical systems and glass and a thousand other things, but until they are brought together to form a system, they represent no more than a collection of unrelated objects. In concert, organized in a unique way to offer mutual physical support and to define space, they create not only a building but also an entire environment within which various activities can take place. All that is added once they become joined in a common purpose, cooperating to achieve that purpose, and thus are part of the system. The whole is truly greater than the sum of its parts.

From a purely pragmatic point of view, nothing else would logically be expected. Nature is an extremely efficient, though often apparently random, system. It tends toward doing things in the most efficient way possible. If there were no positive advantage to a specific type of form or behavior, the natural order of things would quickly eliminate that form or behavior. It is only because there is an advantage to being part of a system that the system exists in the first place. It represents a higher degree of efficiency and order. Without that added advantage (the synergy of being in a system), there would be no reason for the system to exist. Inefficient systems that do not exhibit any synergy simply fall apart and decay. Why synergy? Because without it, there is no reason for systems to exist in nature in the first place!

2. *All systems are reciprocal.* This is a deceptively simple sounding concept. Basically, reciprocity merely means that what you put out is what you get back. That is, all actions and all efforts move in both directions. In a physical context, what goes up must come down. Pendulums swing both ways. For every action, there is an equal and opposite reaction. In a capitalistic economic system, we say that wealth is created by producing value for the consumer, and the more value that is offered, the greater the wealth that is returned. More directly, economics says that there is no free lunch. There is always a price, direct or indirect, to be paid for anything gained. In politics, we find that election to office is in return for an agreement to carry out the wishes of the electorate and that failure to do so will result in either impeachment or not being reelected. Socially, belligerent behavior is generally met with defiance or hostility, while friendly, loving behavior is generally met by acceptance and a return of affection. Even in the larger structures of the universe, we find reciprocity from the law of conservation of energy to the balance of gravity and acceleration exhibited in the orbital paths of planets. There are, in fact, no cases in the physical universe in which reciprocity is not absolute.

It can be argued that this law is not immutable. The world abounds with cases in which reciprocity does not seem to be true, as in the case of those who cheat in business and become rich at the expense of others or the person who gives all his or her love to another only to be spurned and rejected. These anomalies are only apparent. In the first example, a very good case can be argued for the "victims" of the cheating business as bringing it on themselves by not doing their homework, not looking for the best deal, or trying to get something for nothing (greed results in victimization by the greedy). As for the perpetrators, they may get away with shoddy business practices for a while, but eventually they find that no one will deal with them because of their reputation (with the exception of those looking for a free lunch). They may also run afoul of the law and be brought to task for their behavior. In the second example, it should be noted that an individual offering love freely will receive it in return, though it may not be at this moment in time or from the individual with whom he or she is involved. Alternately, what is being expressed here is not love but attachment, which we shall later see is nothing more than a special form of fear. Fear begets fear, and the reciprocity of the situation will see to it that the person's fear of lack ("I *must* have this person's love or I'll just die!") results in lack (the rejection).

The concept of reciprocity can make life quite easy. Since it is true, we can predict everything from political and economic changes to personal gain simply by examining the behavior of the systems involved in those operations. Perhaps we cannot predict exactly how the reciprocity will take place or the exact details, but the overall results can be predicted.

Why the vague nature of the prediction? It is because the systems are complex systems and, therefore, contain so many elements that an accurate prediction is not always possible. Yet this condition does not negate the truth of the statement that the systems are reciprocal. Consider the economy. It is an incredibly complex system, containing millions of businesses and people and many billions of transactions and interactions. How could we be predictive about the future of the economy? There is an old joke that states that economists were invented to make weathermen look good. Another says that the interesting thing about economists is that wherever there are four economists, there are at least five opinions! With such a variance in point of view and accuracy, how can we say the system is reciprocal? It is because of all these elements in the system. Technically, if we could measure the actions of every single individual and group of individuals in the system along with the behavior of production functions, changes in technology, behavior of weather, political systems, social structures, and changes in attitudes and philosophies of life, we could come up with an accurate statement about what was going to happen to the economy in the future. But this is a practical impossibility. There are too many elements, and their behavior is far too capricious. Yet their behavior exactly and directly determines what goes on in the economy. It is not that systems are not reciprocal; it is just that the systems are too complex to trace the steps back to see what actions are going to result in what reactions. This is known as deterministic randomness. The principle points out that if I start with a given action on the part of some individual or group (an element in the system) and follow the reactions of other elements, I can see how that action led to a given result. It is a deterministic process, like a mathematical formula. Plug in a number, run the formula, and the result is determined. Yet it is also random; if I start with the results and try to work my way back to the cause, the more complex the system, the more difficult the task of finding the cause. In the mathematical example, imagine being given a solution and being asked to find the original conditions that created that answer. There are dozens of possible paths leading to a given solution!

The result is that what is actually cause-and-effect reciprocal behavior appears chaotic and random because we simply cannot trace our way back to the cause. Thus when we find someone who has great success in business, it looks like a simple matter of good luck. But there is no element of luck in the events of the world. There is only cause and effect, no matter how random they may appear. The truth is that in some way, the behavior of such individuals in their business affairs led to the "good fortune." This idea is at the very base of chaos theory, which points out that in every chaotic system there is order and that in every ordered system there is chaos. It was not an accident that we decided to use the letters that we chose for the English alphabet, though many other symbols could have been chosen. It is no accident that we are dependent on internal combustion engines for much of our mobility rather than steam or electricity, though any of these options and many others are possible. It is the position of this book that in truth, there are no accidents. All events are created by other events through the reciprocal behavior of systems, and if enough analysis were applied to the processes operating, this could be demonstrated.

That's a nice concept, but is it practical? If the systems are too complex to determine the reciprocal steps, is it useful as a predictive device or guide to behavior? Actually yes, just as the use of electricity is practical though no one has yet determined exactly what it is. If we know how it behaves, we can use it whether or not we know what it is.

Likewise, if we understand that systems are reciprocal, we can use that information even if we cannot determine exactly how this complex system gets us from one point to another and it appears random. I may not be able to predict the exact result of not servicing a machine, but I can predict the nature of that result. It will be reciprocal and disastrous. The machine may seize up. It may explode. It may catch on fire. Or it may simply have a shorter than predicted useful life. But one thing is certain: Not maintaining it will exact a price of some sort. Building with inferior materials results in weaker buildings. Using power that releases contaminants into the air will have a negative effect on the environment. Destruction of habitat will result in less abundant wildlife and a decrease in the quality of life for all of us. Ignoring good rules of behavior in your life will result in poor health, less wealth, fewer opportunities, and a generally lower level of peace of mind. How it will become manifest in your life is an individual proposition. What will happen in the broader sense is not.

3. *All systems are in the process of achieving and maintaining balance.* In truth, this process of achieving and maintaining balance is the goal of any system. Though the manifestation of that goal differs from one system to another, the goal of every system is to create a given set of conditions and maintain them. Again, this represents the raison d'etre of the system. We can define the nature of a specific system in terms of what its goal is and then determine the kinds of behavior to expect from it in terms of what must be done to reach that goal. For instance, for humans, the goal of the organism (in a purely physical light) is to survive, thrive, and reproduce. This is equally true of every organism on the planet. Yet how each different species goes about the process of reaching this goal is determined by its niche in the ecological fabric and how it interacts with every other organism in that ecosystem. Even within a specific species there are different ways of achieving the goal. An eagle living in the Canadian Rockies would hunt different game and build its nest in a different location within the landscape from an eagle dwelling in the southern United States or along the seacoast of Nova Scotia. Humans, too, differ in the manner in which they go about the process of surviving, thriving, and reproducing. We have many different ways of doing this in terms of the jobs we hold, the social and political arrangements we create, and the behavior we exhibit in dealing with others. Yet we are still about the process of creating and maintaining balance. This is true not only of each system but also of the larger systems to which they belong and the subsystems of which they are constituted. All elements, all systems and subsystems, are interacting with other elements and subsystems to achieve their individual goals and reach a state of balance through the achieving of their goals. It is the individual definition of their goal that defines what balance looks like for any specific system. One person views balance as a wife, two kids, two cars in the garage, and a good job that pays $120,000 per year. To another, balance is the freedom to pursue his or her craft or field of study and be left alone, free of the social entanglements of society. Still a third views balance as a matter of integrating a career, child rearing, running a household, and writing novels. Each is working for balance. Yet they must constantly adjust to changes in the greater system of which they are a part, which is why, once achieved, balance must be maintained. They must maintain that balance by constantly adjusting to the needs of others, the changes in business or needs of their children, writer's block, rising cost of living, changing political conditions, and a whole host of other factors imposed by that larger system. Thus every system must not only achieve the balance but also maintain it, by continually making adjustments to achieve its goal. This is why we call the system a dynamic system rather than a static one.

The adjustment takes place automatically for routine changes, such as the alarm clock going off or driving to work and adjusting to the driving decisions of others on the road. Other adjustments are unusual and not so automatic, as in the case of a firm reacting to a competitor who announces a new product or an engineer who discovers a new approach to solving some practical problem. Yet we do react and adjust to bring ourselves back into balance and then react again when the next change occurs in our lives, and this process is continual.

Notice the reactions of the ecosystem to changes in how the subsystems change. There is, for instance, a relationship between the number of foxes and hares living in any natural environment that reflects this systemic change. If the rabbit population rises, the number of foxes rises as well, reflecting the greater availability of the food source and higher survivability among fox kits. When the fox population becomes so large that overhunting occurs and the population of rabbits begins to decline, there is a decrease in the number of kits that survive, and the fox population decreases in response to a smaller supply of game animals to feed the population.

In economics, when there is an increase in demand for some product or in the economy as a whole, manufacturers respond to the shortage by producing more, spurred on by the higher prices consumers are willing to pay as a result of the shortage. When the production rises to the point at which it can satisfy consumer demand, the price drops (no more shortages) and may actually fall below previous levels, resulting in a reduced desire on the part of producers to supply the product. Each of these examples reflects the fact that dynamic systems are constantly adjusting to external circumstances to reestablish equilibrium, achieve their ongoing goal, and be in balance.

All this information about the systemic nature of the universe may be quite fascinating, but what does it have to do with ethics? This subject will be discussed later, when we will tie together the various theories on ethics and the definition presented at the beginning of the book to arrive at what it is hoped will be a useful, pragmatic scheme for determining ethical behavior.

COUNTERPOINT AND APPLICATION

The obvious flaw in the systemic model lies in the apparent observable exceptions that exist to its functioning. These anomalies lie in the realm of the chaotic element of the systems we inhabit and of which we are constructed, and this has been discussed in the text. As for the universal laws, seeing the truth of them may require patience and a paradigmatic shift if you are not used to seeing the world in this light. These laws do exist and they are quite immutable. They are the reason that no one is able to get away with anything and that the system itself is able to survive. If they were not true, we would have no structure. They are their own defense. For those of you still having discomfort with this approach, that's fine. All that is required is observation and allowing for the possibility that there is truth in the information in this chapter. The rest will take care of itself. The following exercises are a great help in starting this process.

Applying this information to technology is fairly simple. As we use and develop technology, consider how the system will be changed, the ways it will probably have to shift to go back to balance (there is no way to predict them all because of the complexity of the system), and how the technology is synergistic. Most importantly, consider how your interaction with the technology in question will come home to roost. What are the reciprocal results of your actions? That becomes the prime directive in terms of the effect of the change not only on yourself but also on other people.

Secondly, understanding the world in terms of systemics will greatly increase our understanding of why things happen as they do and aid in predicting the nature of future events, though the actual content may not be so easily understood. This is a karmic process in the truest sense of the term in that action begets reaction and the reaction is in kind. It is as simple as that, as I believe you will begin to see.

EXERCISES

1. Wherever you are at this moment, look up and look around. Count the systems that you can see just from where you are sitting. Remember the definition: A system is nothing more than a group of elements that interact to reach a goal. How about the book in your hands or the reading light you are using? Is there a computer in the room? A window? What else do you see? Begin to realize that everything is both a system in itself and part of larger systems. List your observations. A word of caution here: If you try to carry this process to its logical conclusion, you'll find yourself counting systems until you reach a state of exhaustion.
2. If the three universal laws of systemics are truly universal, then they must apply to everything from nature to the most contrived of human devices. Consider the ecosystem, the economic system, and the college you attend. In each case, see if you can state how these systems are synergistic, how they are reciprocal, and how they behave in order to achieve a balanced maintenance of their purpose.
3. According to the theory of systemics, all systems are dynamic. That is, the systems that exist for any period of time change to react to changes in the environment (the larger system of which they are a part). To develop a sense of how this is true, consider the dynamic within your own family. How is it changing on an ongoing basis to reestablish balance? Now consider some system that you are aware of that no longer exists. It could be a defunct company or a piece of old machinery or perhaps a friendship that has gone by the wayside. How did these systems come to their demise? Which of the behavioral laws of systems were violated? Did they lose their purpose? Did they cease to be synergistic? Write an account of their demise.
4. To establish what this has to do with ethics, consider the definition that we have been using for ethics: Ethics is doing what

works. How does information about the systemic nature of the universe help us to understand this process? Put your thoughts on this in writing and let yourself speculate. There are no exactly right answers here.

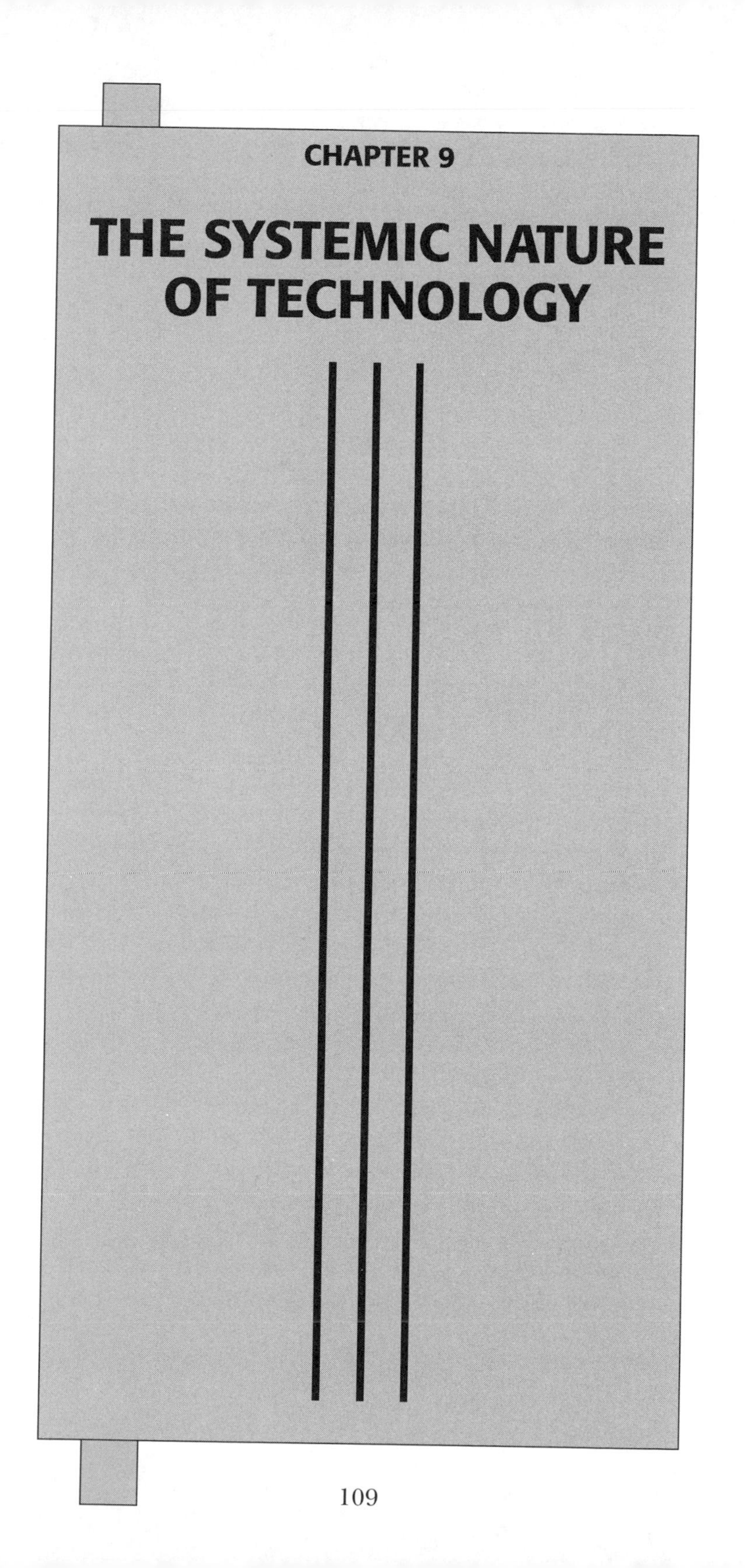

CHAPTER 9

THE SYSTEMIC NATURE OF TECHNOLOGY

INTRODUCTION

After reading the previous chapter, it should come as no surprise that technology is made up of systems. Since virtually everything we encounter in the objective world is systemic, this should be no less true of technological devices than of anything else. Yet technology represents a particular class of systems. In the case of technology, the systems are purposively developed and constructed, their intended goals are predetermined, and the elements that are to make up those systems are limited from the beginning, though they may be changed through time. In other words, with technology, we are dealing with artificial systems designed to produce specific, predictable, and usually repetitive results for our benefit.

TECHNOLOGICAL SYSTEMS

At their simplest, technological systems consist of machines and machine processes. They are designed to carry out a particular set of tasks and nothing more. Admittedly, at times these tasks may take a wide variety of forms, yet they are still very specific tasks. Consider, for example, the computer. Though a wide range of material may be produced using a computer, ranging from business reports to airfoil designs to this book, the computer is still designed to do specific tasks. It is designed to calculate electronically in binary notation using Boolean algebraic principles and generate various arrays of pictorial graphics or servo-mechanical results on the basis of those calculations. Put that way, it seems rather mundane, in spite of the range of results that we can receive from the machine.

Other machines perform other specific functions that are equally simple or complex. If we take a closer look at the nature of machines and nonmechanical technology, we will find the principles of systemics firmly entrenched in their construction. The development of each begins with a goal and then proceeds with a process of determining what is needed to achieve that goal in terms of elements and the interactions among those elements. Finally, the system is constructed and tested to be certain that it carries out the necessary processes to achieve the original goal. Input-process-output-feedback in action is the way technology comes into existence.

Yet to fully understand the impact of our ability to use the systemic process to create technology, we must take into account the complexity of physical and human systems and the resulting uncertainty that results from that complexity. If there were nothing but definable outcomes to all technology, then any created technological device would yield its intended results (reach its goal) and nothing more. Yet this is not always, indeed is seldom, the case. This is because every system is an element in a larger system, and the larger system is in turn part of still larger systems. It is impossible, given the level of complexity the total human system incorporates and the natural system beyond, to predetermine what will result from a technology once we have constructed it. In other words, there are unintended consequences. Predictability is temporary at best as regards technology.

SYSTEMIC RELATIONSHIP

We know that given the appropriate inputs and relationships, a system will create a definable result. But it simply does not stop there. That result alters other systems. It may totally change the nature of the larger system of which this machine is a part. It leads to still other shifts and adjustments by every part of every system with which that machine comes in contact. Like ripples radiating from a pebble being dropped in a pool, it spreads throughout its environment and subtly changes everything. There is no such thing as a machine that only creates the results intended. Too much else can happen in the process to be fully realized. We can only wait and see. What we do know and what we can count on is that there will be a reciprocal result from this process and that, eventually, all of the systems involved, from the smallest to the largest, will seek and maintain balance.

Consider a medium-size software company in the field of developing multimedia products for the market. Given that the people in the firm are qualified for the job, that the creative individuals charged with defining new products and the programmers/artists are effective at translating those product ideas into reality, the result will be a product ready for the market. For the sake of argument, let's assume that it is a remarkably intuitive three-dimensional interface for navigating databases. Let us further stipulate that the employees are able to successfully market this product, producing revenues for the firm. It would appear that the system has met all the criteria for achieving its goal. It may further satisfy other company goals, such as improved position of the firm in the market, fame for the principles, and the opportunity to expand the business. But the results of this system's machinations to achieve its goals do not stop there. The market itself may be affected by the new approach to database manipulation created by the product. If it is truly innovative and intuitive (thus easily learned), it may become the standard, resulting in the demise of less easily manipulated interfaces or, at a minimum, forcing other producers to alter their products. It may also bring about new levels of efficiency among customers who use the product, allowing them to assimilate, manipulate, and regurgitate data at a greatly increased rate, producing both more rapid and accurate decision making by the users and the proliferation of data storage. This affects still other firms of all kinds and could result in the demise of some businesses and the rise of others. If you feel that sounds rather farfetched, consider the rise of firms such as IBM or Apple as they developed new products. Consider the rather obvious example of Microsoft, which has become a standard in the software industry, certainly in terms of the Graphical User Interface, though arguably that is as much the result of Apple as it is Microsoft. The point is that the original purpose of any technological development does not define the sum total of the results of creation of that technological development. No technology exists in a vacuum, any more than a person exists exclusive of the rest of humanity. All elements in a system are connected, and that system is connected with all other systems in turn.

Electromechanical devices have very specifically determined subsystems within them, and we are able to predict what each of those subsystems will do under a given set of circumstances. All of the circuitry and gears of a mechanical device will behave in predictable ways according to natural laws, and therefore we can design and create the new mechanical systems to perform specific tasks. The synergy is obvious in that the device does useful work, does it better than if it did not exist, and does it in a new way. The reciprocity stems from the fact that the elements that are put into the system and whose relationships are defined by the construction of the system result in new behavior that is exactly in accord with the elements put together. It is in the balance that we find our predictive problem.

Control systems are included in our mechanical devices to create and maintain balance. Mechanical governors, cybernetic control mechanisms, gyroscopes, and cooling systems are all examples of automatic methods of seeing to it that machinery and equipment do not run amok and self-destruct. Yet even if a system is operating perfectly, it creates unexpected results.

Consider a machine designed to harvest some agricultural product. Suppose you invent a superior lettuce harvester that takes the place of hand picking. It is faster, it is much more predictable and uniform than a worker, and it can even discern which heads should be picked and which should not. If it performs perfectly, it replaces people with machines, and the harvest is brought in more efficiently and more economically. All of this may represent the original goal of the picking machine when the engineers decided to develop it in the first place. It sounds like a prime example of an artificial mechanical system (device) that successfully fulfills its intended purpose. It takes little imagination to discern, however, that the invention, development, and use of this machine creates other secondary and tertiary effects as well.

Every action has a reaction. That does not happen just in terms of billiard balls and rockets. It happens in economics and politics and technology. In truth, it happens to all of those and a myriad of other fields of human activity simultaneously, no matter what the original trigger cause may be. We build a machine to pick lettuce. The results are not only a more efficient harvest but also, perhaps, lower prices for wholesalers and retailers who purchase the crop. Additionally, it results in a reduction in the need for migrant workers who have been picking the crop and an increase in acreage devoted to lettuce as the demand rises because of lower prices. This leads to less of other crops being produced and made available because of this shift in agricultural resources, the development of political action groups among migrant farm workers coupled with calls for welfare reform in the affected states, and on, and on, and on. All of this happens because someone decides to find a better way to pick lettuce.

The lettuce-picking machine is just a single example. It in no way argues that we should not build lettuce pickers because of the other issues that arise. It suggests, rather, that the systemic nature of technology is no different from the systemic nature of the other elements in the ecosystem. The change in lettuce-picking technology affects every aspect of our lives and our way of life merely by existing. The point is not to avoid technology but rather to realize *that we cannot always predict accurately the results of that technological process* and that to predict the sum total of the results of a technological process is virtually impossible, given its connections to the rest of the system. There are unexpected costs that will arise with the benefits from every technological advance.

COUNTERPOINT AND APPLICATION

It is hoped that reading this chapter has served to clarify much of the material in the last chapter, particularly the material in the Counterpoint and Application section. Primary criticisms of the ideas in this chapter usually come in the form of 'So what?' or "If I can't predict what's going to happen and all technology has undeterminable effects on someone someplace, how can I use this to solve ethical dilemmas or determine right action?" These are valid concerns.

The answer is that although we cannot predict the exact nature of technological change, we can use our knowledge of systemics to approximate the kind of results we can expect and to minimize the effects that we cannot predict. Once the changes become manifest and we find that the unexpected has occurred, we are not blindsided by not expecting unusual results and are better able to understand their cause.

If we invent a new method of mass farming, one that increases yield while decreasing environmental concerns, we expect both of those events to occur. We may not anticipate the negative effect these results will have on small farmers or on the manufacturer of alternative equipment that suddenly becomes obsolete. Yet when those results occur, when equipment manufacturers start legal actions to protect their investments or challenge patents and when small farmers find themselves hard pressed to compete, we will be less surprised and in a greater position to respond. In the first instance, we may find mutual ground for economic growth with other manufacturers by licensing the production of our new device. In the case of small farmers, they may discover that there are higher profits in switching to more exotic crops or supplementing income by adding alternative land use (e.g., harvest festivals open to the public or living museums of traditional agricultural techniques in off-seasons). It must be remembered that when one technology replaces another, it results in the demise of the older technology and that this is not necessarily a bad thing. There is definite short-term hardship for those losing their businesses or their jobs, but in the broader sense, it is a healthy shift of resources from a less efficient methodology to a more efficient methodology that benefits all concerned. Workers can be retrained. Companies find themselves forced to become creative in finding new uses for their products or modifying them to compete with the new technology. They may sell current holdings to some other entrepreneur and enter new businesses that bring even greater success. In a modern world in which the rate of change is so rapid, those who stand still are actually moving backward; innovation can often prod people into further progress.

EXERCISES

1. Technology is a relatively simple area of human experience in which to look for systemic effects, because we have produced these systemic elements consciously. A principle or idea may

come to someone by accident, but the creation of the mechanical device or the technological methodology is created on purpose. It is defined and structured with specific goals in mind. Think about this as you pick any three of the following technological systems and determine (a) what the original intended goal and expected effect of that technology was and (b) what the observable results to our lives has been as a result of that technology's use.

a. Agriculture
b. Radio
c. Television
d. Computers
e. Assembly lines
f. Automobiles
g. Inoculations
h. Plows

2. All inventions and technological devices have unintended consequences because of the complexity of the overall system. Speculate on the probable results of the development of an automobile that hovers by the employment of antigravity technology. To limit the range of consequences, assume that in other ways the vehicle is similar to modern automobiles, that it operates on electrical batteries, that its altitude has a maximum range of fifty feet above the ground, and that it is comparably priced with modern vehicles as well. Look not only for obvious changes (primary ones) but also for secondary and tertiary changes.

3. Consider the pencil sharpener as a technology. What are the subsystems that make up this device? What is a system of which the pencil sharpener is a part? What is a system of which *that* system is a part?

4. Assume that human cloning becomes commonplace at some time in the near future. What ethical issues do you see arising from that technology in (a) the religious community, (b) society, (c) the political system, and (d) the family? Based on these issues, what restrictions would you personally suggest be attached to cloning?

5. Having completed exercise 4, consider your suggested restrictions. What do they say about your own paradigm regarding religion, society, politics, and family? What do your answers tell you about your own learned ethics? Write specific statements of ethical principle based on your reaction to cloning. Try not to judge these statements as either right or wrong.

The purpose of the exercise is to take a look at your own belief systems and see whether they are in line with your conscious understanding of what you consider to work.

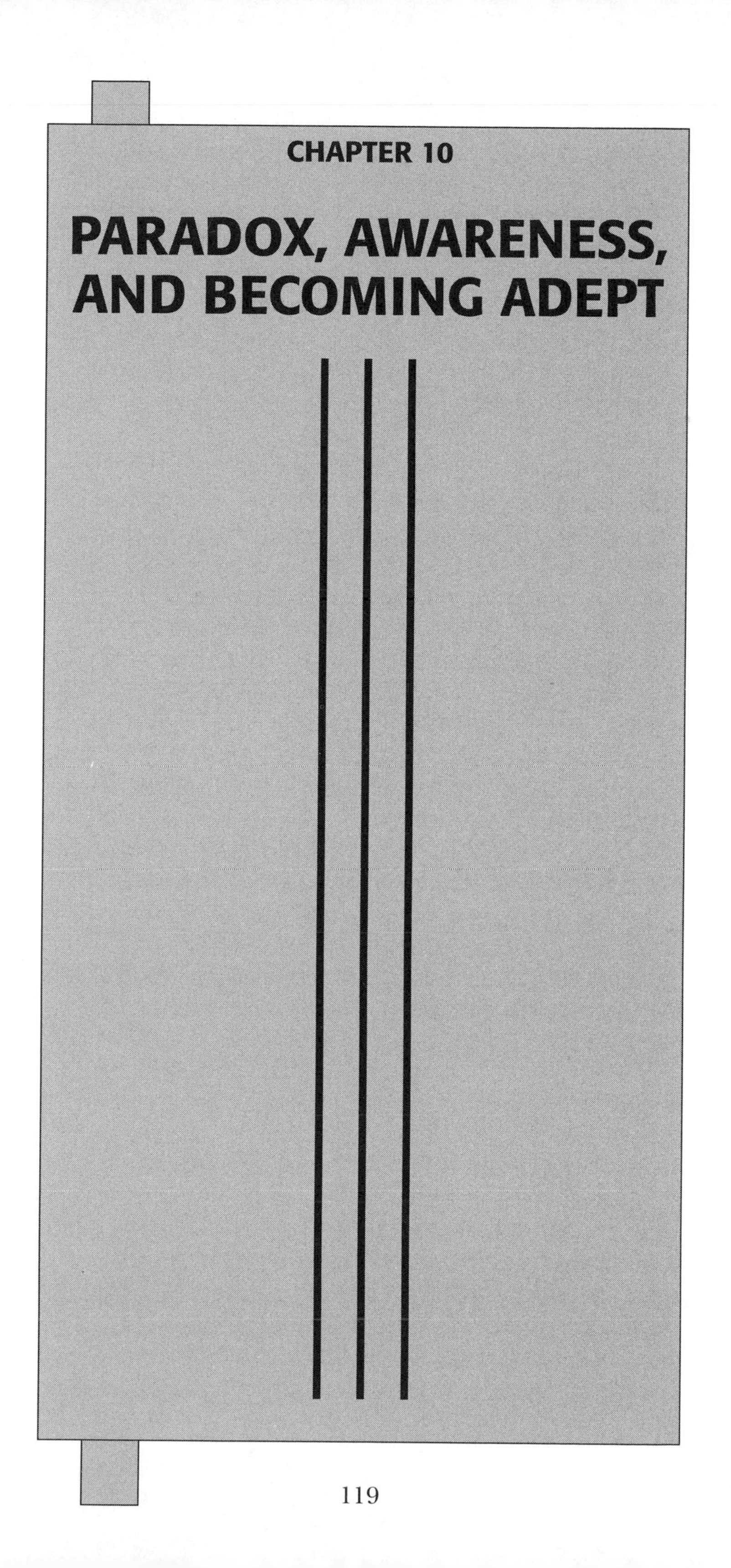

CHAPTER 10

PARADOX, AWARENESS, AND BECOMING ADEPT

INTRODUCTION

We have looked at a great deal of disparate information in the previous chapters, investigating everything from the nature of reality itself to the workings of systems theory. Does any of this information bring us close to understanding how to actually behave if we wish to be ethical? If we accept the stated definition of ethics, is there anything in the previous chapters that supports actual practical methods of behavior that allow us to, in fact, do what works? On first inspection, the topics may not immediately appear to be related, either to themselves or to a practical method of living, an ethical *modus operandi*. But quite the opposite is true.

In the information on the nature of experienced reality, we find that it is individual and not necessarily dependent on the reality that others perceive. This being the case, turning to others for definitive lists of dos and don'ts concerning ethical behavior is not entirely adequate. Though in many cases we will find the opinions and behavior of others to be good guides as to what works, in other cases, we will find that what they have taught us is not accurate. It is still our individual experience that determines our understanding of the world; hence we have individual realities.

In the case of the behaviorists, the information seems to be good information, but again, our own experience is that though this may be the appropriate way to behave, people do not always behave appropriately or consistently. In fact, many people seem to be quite content to dwell in the lower regions of behavioral schema. They are quite content basing their decisions on relatively selfish motivations that do not include much concern for the well-being of others or, if recent trends in business are any indication, concern for their own long-term well-being either. They may even rationalize their actions, seeing their decisions as being in the best interest of everyone, even though it creates pain and misery for some. We have to search far and wide to find anyone who even begins to approach a universal ethic, much less transcendental ethics. This information may be a very nice set of concepts that describe how people *ought* to act, but it hardly reflects the reality of how people *will* react to the opportunities and needs of life. There doesn't seem to be much help here, either.

Finally, the discussions of systemics and chaos theory were perhaps enlightening and may have even struck a chord with the reader in terms of personal observation about the way the world is put together. Yet even if we accept the concepts of synergy, reciprocity, and balance as true representatives of universal behavior, how does it help us? The chapter is quite clear about our ability to understand what goes on. This is a horrendously complex system in which we exist, so much so that being able to understand it, much less manipulate it, seems like an impossible chore. With so many variables in the system to consider, how can we know what behavior works in this context at all? Even if I strive to behave correctly, how can I know I won't inadvertently injure or otherwise cause harm to a fellow human being? And does this extend to other creatures on the planet? How far must I take my attempts to be ethical? Am I to starve myself so that I don't hurt some animal or plant? Isn't that impractical, even if I believe it to be true? What about all the bacteria in my body? Doesn't my immune system automatically fight them? It all seems quite hopeless. Yet it is not. It all fits together quite nicely, in fact. Why all the confusion, and why is all this so evidently paradoxical? Consider the following understanding of ethics.

ETHICS AND PARADOX

Accept for the moment that being ethical is a matter of doing what works. Accept also that this only helps us if we know what works. Now take another look at the information with which we have to work. If the systemic theory is correct, and it appears to be so from all personal and general observation, then all systems are reciprocal. That is, what you put into a system is what you get out. Its effectiveness in doing a job depends on how well focused you are (creating the synergy), but for certain, the effort and energy you put into a system is identical to the sum total of the results. Focused effort creates focused results, into which all the energy goes to achieve the goal of that action. Unfocused energy creates scattered or negative results because that energy is not all going toward the creation of the intended result. Yet result there will be. Put another way, you can't plant corn and get a field of wheat. If you want wheat, you'd better plant wheat. Likewise, if you want happiness in your life, you can't sow unhappiness for others.

In fact, this is true of all action. Whatever action you take creates an equally pleasant or unpleasant reaction for yourself. Note how true this really is. How often have you attempted to gain by deceiving others? How often have you been even slightly dishonest about your motives or your actions? And how often as a result of this have you found out later that the other person or persons involved were also devious to some degree, hiding their true motivations and feelings from you? This is equally true for personal relationships, professional relationships, business deals, and how we interact with other nations. It is a universal truth. What you put in is what you get out.

There's a very easy way to check this. Be very careful here. This could be a painful process for you if you've been hiding from the truth of this concept for very long. *You can always know a person's motivations and original intent by what the results of their actions are.* Know what went in by looking at what the results were. If it worked out well, their motivations were positive. If it didn't, they were about the process of attempting to gain at someone else's expense. This is universal. However, in considering this idea there are two caveats.

The first confusion stems from our belief systems (attitudes, judgments, etc.) about what is a positive and what is a negative outcome. At times, what appears to be quite terrible is in fact a wonderful thing. We view it as terrible because with the limited amount of information available to us, we do not see how it is to our advantage. Later, we may be thankful for events in our life that initially look horrendous. Have you ever lost a job only to be later thankful because it opened up opportunities you didn't know you had? Did you ever lose a lover only to find that there was another, even more wonderful person out there waiting for you? Perhaps all you needed to do was make room in your life for that other person. Why should this surprise us? We have already discussed the fact that the system is too complicated to be really understood. Perhaps finding out what works is also a matter of eliminating things that don't work, and that may need to be by trial-and-error discovery. Failures are often just steps to success. *Of course*, we would prefer to go directly to a winning solution, but sometimes there is some learning necessary along the way. The point is that just because something is viewed as painful or detrimental does not necessarily mean it is bad. At times, it is best to suspend judgment on results until they are better understood.

The second caveat concerns our awareness of our own motives. At times, we simply do not understand the selfish motivations we have for our actions. By definition, selfish motivations, rather than actions that will be of benefit to us in the long run, are generated by fear. We couch selfishness in altruistic terms or pretend we are behaving in one way when in fact we are behaving in another. Consider, for example, a woman who is deeply in love with a man. She virtually worships the ground he walks on. She would do anything for him. She showers him with gifts, wants to spend every waking moment with him, and centers her life on him. She is, obviously, in love. What happens if he leaves her? Of course, she falls apart. Her life is miserable. She feels extreme pain and doesn't want to go on. She may even choose to not go on. Yet if loving is such a positive act, then why is it that he has left her? Isn't that in opposition to the law of reciprocity? Shouldn't he love her back? It sounds like love hurts!

This apparent paradox stems from the fact that such a person, who has lost his or her lover, is confusing love, a positive emotion in which the welfare of the object of the affection is the important issue, with attachment, in which the fear of losing this wonderful person is the motivation for the behavior. If you are more interested in your own well-being than in the happiness of the other person, you are attached. If you want wealth because you are afraid to live without it, you are attached. If you want success because you are afraid that you are a failure and nothing (in your own eyes and the eyes of others) without it, you are attached. You are acting out of fear, and fear is the exact opposite of love. Real love involves a genuine desire for the well-being of others; it is never attached. Attachment is fear, and there are many of these in our lives. It is one of those paradoxes of truth that the way to have what you want is to not want it. If you *have* to have things to feel whole, then those things are in control of your life. If you feel *drawn* to have them, they are probably expressions of your true self and to pursue them is a matter of self-expression rather than placating fear. Some people want high position for the power it brings and the resulting feeling of control. Others seek high position because it is viewed as a method by which they can better affect the lives of those around them in positive ways. Both types may be able to achieve high position and garner personal power, but only the latter will find the experience enlivening. The former will fail to receive any joy at all from the quest.

THE PARADOX RESOLVED

The same is true of any other pursuit. What are your motivations? What are the reasons for wanting whatever it is that you want? If it is an expression of who and what you really are (self-actualization), then it will be to your benefit and to the benefit of others. The only gift you have to offer others is yourself anyway. Anything else is a lie based on fear. Do you modify your behavior to impress others, or do you honestly present who you really are? Do you pretend to like certain things or to think a certain way so that your boss will be impressed or your date will be attracted to you, or do you just let everyone know what you see as truth? Any deception is an expression of fear, and fear is the killer of truth, integrity, self-esteem, honesty, and ethical behavior. It is, in fact, the only reason anyone is ever purposely (whether consciously or unconsciously) unethical. Peace of mind, the prize one obtains from an ethical existence, can only be obtained through freedom from fear. It is only realized by behaving in ways that reflect the perfection of the universe, and that means loving yourself and loving others.

When all is said and done, it is more the motivations of our actions than the actions themselves that create the results in kind. That does not mean that if someone does something truly horrible for a mistaken reason that the results will be wonderful. It means that if your motivation is based on false information, the results are a lesson learned rather than pain and misery. You still get back what you put in. None of us will always know exactly what the perfect action to take will be, but we can learn from our mistakes rather than pretend the reciprocity isn't there. We can remember that it does no good to fight city hall when city hall is defined as the entire physical universe. We can all begin to realize and learn from the results of our behavior.

This is not new information. Every major religion in existence today includes the concept. Every major ethical philosophy has at its core the desire to determine "right" action. Every legend and myth we encounter is basically a lesson in reciprocity. If it weren't, it would not long survive and would have long since fallen away.

The truth is that when we become aware of the real nature of the universe, we realize that people's motivation as well as their behavior determines the outcome of their actions. It is a matter of the heart, not of the head, and in each act, there is a learning taking place that more closely defines the truth of our motivations. Being adept is nothing more than realizing what our motivations are and knowing what motivations to express. The paradox is solved. Ethical thought leads to ethical action, and this in turn leads to a life that works.

COUNTERPOINT AND APPLICATION

In order to accept the premise of this chapter, that reciprocity is absolute and that all actions come home to roost, we must account for all of the instances in life when the opposite seems to be true. We must deal with the experiences in our own lives when people seem to "get away with it" or are punished unduly. We must, in fact, deal with the paradoxes. There are so many apparent cases of injustice in the world, so much misery experienced by seemingly innocent people (victims), that it becomes very difficult to embrace the idea of reciprocity.

This in itself becomes a paradox, and as will be remembered, paradoxes exist simply because we do not fully understand the nature of a situation. In fact, if there is a situation in which you are having trouble understanding the content, consider changing the context. That is, consider the possibility that the problem is not with the ambiguity of the situation so much as your own perspective on that situation. You might begin to think of the situation as containing not problems to be solved but rather opportunities for learning the truth.

Be aware that our lives are opportunities to learn. We are in school here, every minute of the day and night. We are constantly surrounded by opportunities to increase our understanding of the truth of reality and to sharpen our abilities to deal with our existence in effective ways. It is very much a school of trial and error, though we are given tremendous resources to help us along the way, such as the entire content of our cultures and all of the written and recorded wisdom of others that provide us with what others have found and how they have constructed their paradigms. Yet these are only guidelines, suggestions for us to explore as various alternative avenues to discover successful living. It is a very Taoist approach in that it is a journey that is a destination in itself. We are here to learn what is true and what is real. In the process, we are free to believe whatever we wish, but depending on our beliefs (our paradigms), we have varying degrees of success. Wrong paradigms simply do not work. Correct paradigms do. If we believe that one and one is three, for instance, we will soon find there is great pain and misery in that belief; our choice is to cling to it (probably with disastrous results in finance, handling mathematics, holding a job, and so forth) or to change our minds and think of one and one as some other quantity, say, two. Such a shift would immediately clear up a whole class of difficulties within our lives.

This is what is happening when we find ourselves facing paradoxes. We have simply found a situation in which our paradigm does not lead to the results we are expecting. The choice is to either cling to the current belief system or consider alternative views of reality. With a shift in paradigm and accompanying greater success in life, we have learned. There is no difference conceptually between the paradigm shift of someone who suddenly becomes aware of the reciprocal nature of events and someone who learns to do math. Each results in new learning, new tools for living, and higher levels of success.

As a method of applying this concept, think about the beliefs you have concerning technology. For instance, most people whom I have encountered feel that technology always brings problems with its benefits. Every new device creates not only positive effects but also negative ones. We have discussed this earlier. Perhaps if we were to begin to consider the possibility that this does not necessarily follow automatically, we will open ourselves to designing technologies that do not affect us negatively or, alternatively, begin to see how seemingly negative effects may be positive in the long run. Just because some result is unpleasant does not necessarily mean that it is negative. Childbirth can be quite painful or it can be relatively painless, depending on the individual, but in either case, the results are generally seen as wonderful for both the participating mother and the child, not to mention ancillary participants, such as the father, family, and friends. We have the choice of experiencing an event such as childbirth or the birth of a new technology as either painful or pleasurable. Which would you choose?

EXERCISES

1. Consider some of the myths of your childhood, particularly the standard ones that are part of the culture, like the stories of Little Red Riding Hood or Goldilocks and the Three Bears. How are these myths the exposition of ethical behavior? How do they indicate either synergy, reciprocity, or balance?
2. Now consider a modern soap opera on television. There are several conditions that we know exist in these dramas, including that things go well only momentarily and that whatever happens on Friday is going to be left unresolved until Monday to get the viewer to tune in again next week. However, what is it about these melodramas that makes them so appealing and in many cases so realistic? How are they reflections of reciprocity, and how do the characters reap what they sow?
3. As stated previously, those elements in our lives that we are attached to are the ones that are not held in love but rather in fear. List four people, four possessions, and four ideas to which you feel attachment (i.e., if you were to lose them, you would be miserably unhappy). Now analyze the nature of the fear that creates the attachment. How might you still feel the close connection to these listed elements of your life yet do so without fear?
4. Examine four important events in your life such that two were positive events and two were negative events. Assuming that the content of your life is the result of your own behavior and not the responsibility of someone else, determine how each of these four events was created by your actions. What does this tell you about how to behave in the future in a manner that works for you?
5. Put the concepts of synergy, reciprocity, and balance to the test. Develop some minor action that you can take that is of benefit to another and carry out this action. In other words, do something positive for another human being or group of

human beings and watch the results. Where is the synergy in the process? Where is the reciprocity in the process? How does the action and results (for everyone, including yourself) indicate balance?

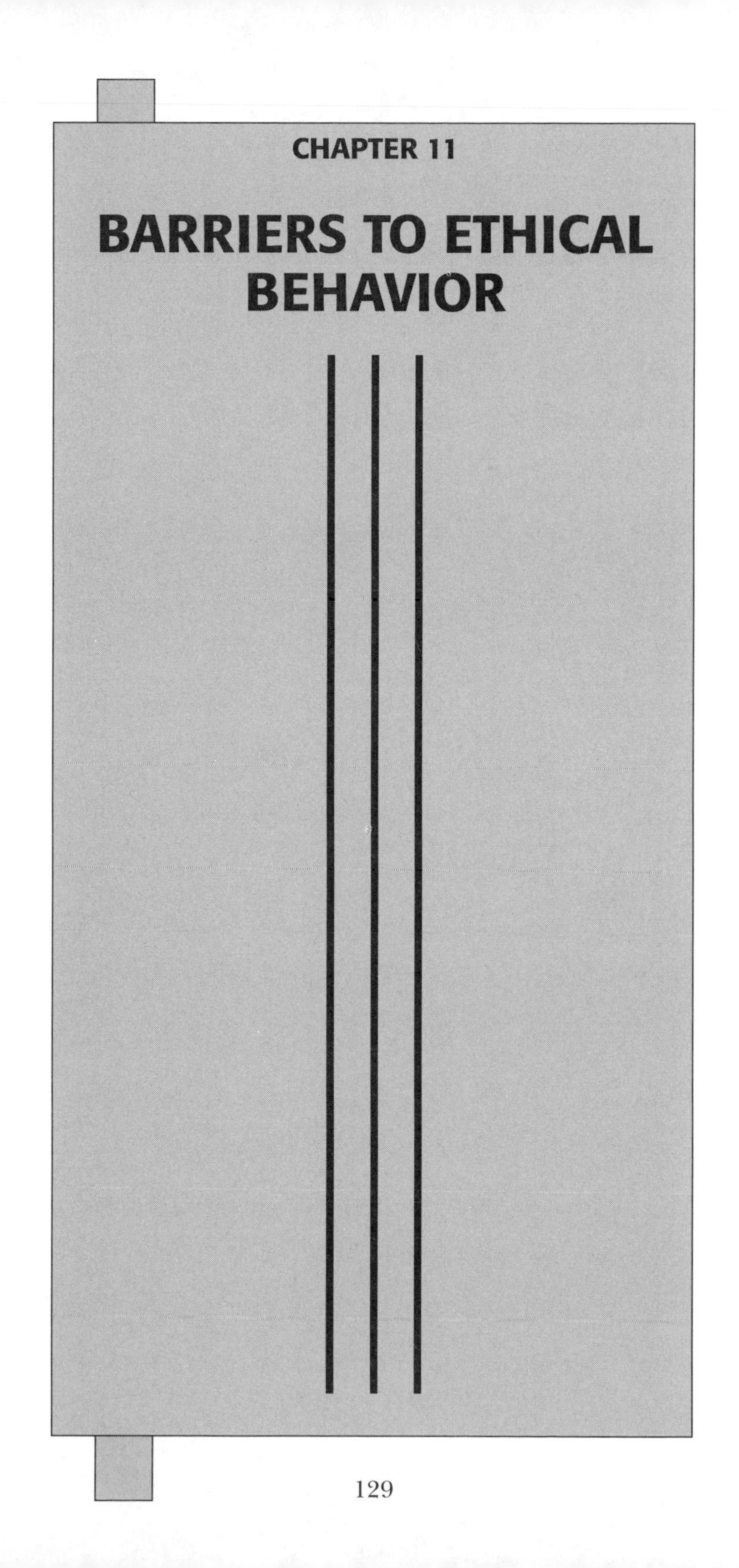

CHAPTER 11

BARRIERS TO ETHICAL BEHAVIOR

INTRODUCTION

Virtually all individuals can remember experiencing situations in which they tried their best to create a given positive result from their behavior and ended up with some other undesirable result instead. It appears that no matter what they consciously attempt to create, it goes sour. Where are the rewards of their efforts? Why is it so difficult to create what they desire in life, and what keeps all this wonderful, positive benefit from coming about? Is anyone really that "good"?

To begin with, "good" is a relative term, a value judgment like right and wrong, or moral and immoral. If we define good as anything that creates peace of mind, then we can explore what it is that keeps us from achieving that peace of mind. Semantics aside, the questions remain: Are we capable of achieving peace of mind, and are we keeping ourselves from achieving such a state? Actually, the answer is that peace of mind can be achieved, if we simply realize how we continue to block our own happiness, even when we know what it is that will create the happiness.

FEAR AS MOTIVATION

The truth is simply that when we act in ways that are unethical or are detrimental to our well-being, we are afraid. Fear is the single barrier that separates us from achieving peace of mind, whether we are speaking technologically or in any other terms. It is the second most powerful motivator of behavior, and since there is only one other, love, it would appear that *the source of our personal misery is fear*. This is true whether the fear directly causes us pain or our behavior as a result of fear causes us pain through the process of reciprocity. In either case, fear fuels our hatred and our greed, our violence and our apathy. It forces us to protect ourselves from others and to modify our behavior, shifting from our natural tendency toward cooperation, compassion, and consideration for the other person to other behaviors that harm not only ourselves but others as well. And it is in the reciprocity of the natural system that this fear behavior creates our misery.

Consider individuals who are raised to believe that it is a ruthless world in which everyone is out to take what others have and that you must protect yourself and your property at all costs. If that is the world paradigm of some individuals, what kind of world are they going to create? What is their behavior going to be? They will be suspicious, distrustful, and avaricious. Their behavior will reflect a desire to take advantage of every situation and every individual, turning every interaction to their own benefit and living at the expense of others. If they are to live in such a world, they will naturally become part of the process just to survive and develop a "bunker mentality" designed to protect themselves at every turn and at the expense of everyone else.

Admittedly this is an extreme example of the principle, yet how often have you felt you must seize an opportunity when you are in a position of advantage over someone else? How often have you used superior knowledge of a situation or superior talents to outwit an opponent, get the better of a competitor, or glean a bit more benefit from interactions with others? If you take a close and honest look at your own personal behavior, you will probably find many examples of this type of behavior. People exhibit their fear in many ways. They may create huge reserves of wealth to protect themselves from possible economic disaster. They may pass up opportunities that require a measure of faith or willingness to take a chance. They may avoid being charitable or giving to others because the other person might take advantage of the generosity. They may fear the other person is lying about his or her needs. They might not open up to another person for fear of ridicule. Or they may hide the truth of their own past for fear of being ostracized or looked down upon.

People fear the unknown, failure, success, revealing the truth, poverty, wealth, and the honesty or dishonesty of others. In this fear, people react to conditions by defending themselves, and in so doing, they create the need to be defended. Reciprocity is absolute. In defending ourselves, we create people in our lives that defend themselves. In lying, we create liars in our lives. By behaving greedily, believing there's not enough to go around, we create a greedy world in which others are out for themselves, are very ungenerous, and will take advantage of us if we let them. As we believe the world to be, we will respond to it, and as we respond to the world, it will respond to us. The fearful conditions of our lives are the result of our own fear, not the nature of the world. Feeling safe in an unsafe world is a matter of realizing that we create the circumstances of our lives as individuals and that fear is met with fear as love is met with love. It is the great paradox that we do not see this, though it is presented to us and proven to us every day of our lives. We choose to be blind to the truth, even when we are made aware of it.

A CYCLE OF REVENGE

Now consider a man who enters a restaurant for lunch. He is happy, he is in balance for the moment, and he has not a care in the world. Yet when he enters the restaurant, a patron leaving bumps into him and just keeps going. Startled and off balance, the man turns and says, "Hey! Why don't you look where you're going?" In return, the patron growls a foul remark and offers a universally recognized gesture with his finger. Our subject considers physical violence, remembers where he is, and enters the restaurant, growling.

When the waiter seats him with a pleasant smile, he asks our subject what he would like to drink before taking the order. "If I wanted a drink, I'd have gone to a bar!" barks the man, still angry over the exchange at the door. "Just bring me some water and leave the menu!" The waiter's expression changes, but he says nothing, curtly hands over the menu, and walks away. Several patrons, having heard the exchange, look over at the seated man, who returns their attention with a frown and a glare. "Can I help you?" he says sarcastically. The other patrons turn away and back to their own conversations.

When the waiter returns, the water is placed on the table none too gently and the waiter says, "Are you ready to order?"

"What's the hurry?" the patron asks, sensing the hostility.

"No hurry, sir. Take your time."

"I'll have the special."

"Certainly, sir," says the waiter, and leaves.

Ten minutes pass. Neither the waiter nor the food appears, and between his own anger and the impatience he is feeling, our subject develops a degree of discomfort in his stomach. He spots the waiter three tables away and calls out loudly, "Can I get some service here?"

The waiter looks up startled, calms himself, and says, "One moment, sir, I have other customers. I'll be with you momentarily."

Now truly enraged, the patron bellows, "What kind of place is this anyway? Why am I being kept waiting? Can't you people get anything right?" His stomach begins to ache, his face is turning red, and he becomes extremely agitated. Two gentlemen at another table actually start to rise.

With perfect poise the waiter says, "Perhaps you would prefer another establishment, sir?"

"You're darn right I would," our subject growls, rises, and walks out of the restaurant. All the way back to the office, he thinks of how angry he is, how enraged and insulted he feels being treated so poorly at the restaurant, and how rude people have become in recent years. At the office, he buys two candy bars and a soda from vending machines and returns to his work. He huddles over his desk, grumbling at his coworkers, and plods angrily through report after report. He has a thoroughly miserable afternoon.

This man's lunch experience totally ruined his day. Not only did he fail to have a nice lunch, he experienced physical pain, felt the anger of the waiter, and will probably not return to that restaurant for some time to come. The interesting thing is that it all seemed to be caused by another patron who bumped him as he entered the door and failed to say excuse me. Everything went downhill from there. The man chose to be angry, and in his anger over the incident, he chose to take out his frustrations on the waiter. The result was that the waiter promptly returned the favor by offering poor service and the man risked bodily harm from other patrons and lost the ability to visit one of his favorite restaurants in the future. And in his own anger, the waiter lost the opportunity of a nice tip. If the manager was watching the exchange, he may have lost the opportunity of his employment. Even the other patrons of the restaurant who stared received a dose of threatening sarcasm as the patron felt the anger and fear rise in him. They all lost; every one of them. And all this occurred because one man bumped into another man coming in the door. Does this make sense? More importantly, does it sound familiar? How often has one negative experience led to another in your life? How did all this happen?

The principle demonstrated in this scenario is clearly one of action-reaction. Our subject went through a familiar cycle that nearly everyone experiences from time to time. It started when he resented being bumped. He simply resented it, and in response, he sought to strike back verbally. Of course, reciprocity being what it is, the person who bumped him, rather than apologizing, fired back a rude retort accompanied by a rude gesture. Now even more resentful, the man entered the restaurant and proceeded to take it out on the first person he encountered, the waiter. The waiter, we may surmise, experienced his own resentment and, though he controled himself, chose to counter with revenge of his own through his attitude and treatment of the customer. Each act of revenge resulted in more misery for the patron. His meal was late, people stared at him, and he had stomach pain, which he blamed on the lack of food; he ended up leaving, again as a last statement of *revenge*.

What causes incidents like this to occur in our lives? Mainly, it is a matter of reacting to events with fear, rethinking the events and blowing them out of proportion, and finally attempting to protect ourselves by seeking revenge. We all do it every day, in a thousand little and sometimes not-so-little ways. Everything from a curt remark to a spouse to an act of road rage can be traced back to some immediate or longstanding fear that results in an irrational desire to protect one's self through revenge. Note that revenge is always violent, whether physical or psychological, and that violence is always the result of fear.

TRIGGER-ANGER-REVENGE

Trigger-anger-revenge (TAR)—this is the sequence of events that lead us into these self-created miseries. The trigger is the point at which we become afraid, for whatever logical or illogical reason; the anger is the point at which we let that fear build to where we feel the need to defend ourselves; and the revenge is the expression of that violence. In every single case, however, the persons who are injured by the revenge are ourselves. Who lost in the exchange in the restaurant? The patron received no meal, the waiter received no tip, and the staring patrons received verbal abuse. In each case, the move to express the anger resulted in a negative reaction. It is a reciprocal process. What you put in is what you get back. Anger begets anger. Revenge begets revenge. What else could we expect?

Yet it seems rather naïve to suggest that people are not going to react to threatening events or get angry. That is extremely unrealistic. Does this mean that we are all doomed to cut our own throats when we are triggered? Fortunately, no, and the reason is that there is a choice at the point of trigger.

The trigger that starts the process, the action or event that we first resent, is the key to this process, and it is there that we can turn the process around from a negative one to a positive one. The place to change the situation is at the source of the fear, and that is when the resentment first arises to trigger the sequence of events.

Resentment is, paradoxically, not a negative experience. It is only what we do with it that creates the negativity. In truth, resentment is unavoidable and valuable to us, because it represents a basic fear with which we have failed to cope. Each resentment a person harbors or experiences is the result of some fear-creating experience with which he or she has not dealt. Since the only way to deal with a fear is to know what it is, the resentment is valuable. It makes us aware that we have unresolved issues. If we wish to deal with our fears, they must be identified and understood. Otherwise, we are helpless in moving from the state of fear to a state of acceptance.

In the normal sequence of events, when a resentment arises and we are triggered, we experience an intensification of the feelings that accompany the fear until we are moved to defend ourselves and the vengeful violence results. Yet this is not the only path available to us. If, at the first sign of resentment, we were to stop and simply work on discovering the source of the fear, we could remove the trigger and thus not fall into the trap of going, kicking and screaming, to revenge. We would, in fact, reduce the number of fears that we have and, in the process, reduce the number of ways we can be triggered into anger.

If we rethink the example of the restaurant patron, we find that his entire experience began because someone bumped into him and did not say excuse me. He resented the fact. This is the trigger that started the sequence of events leading to his misery. Why? People bump into each other all the time, and on some occasions, they do not say excuse me. Why is it resented? For different people there would be different reasons, but generally it will center on the fact that there was bodily contact that was not sought or expected and that it happened in a manner that was physically shocking. He did not expect it.

In a civilized society, we have developed ways of defusing such potentially violent encounters, including the tendency to say excuse me if we happen to inadvertently make physical contact with someone in an inappropriate way. When this apology is not forthcoming, there is a natural reaction. For some people, it could be an immediate physical confrontation; for others, it is nothing more than mild irritation. Our example patron chose to become verbally abusive, for which he received verbal and visual abuse in return. Now he resents the abuse as well, and so it goes. But why did he become angry enough to snap at the man to begin with? Why did he not choose to ignore it or just be mildly irritated? It is the nature of the resentment that creates the reaction.

The alternative reaction would be to realize that the resentment had nothing to do with the other person and to look for the internal cause instead of striking out. For most people, that concept seems quite bizarre at first. Think about it. If the cause of the anger was really the action of the other person, then everyone who ever had that experience would react in the same way. By way of another example, we would have to expect every waiter who was ever growled at by a customer to immediately become slow to give service and curt in return. Yet this is not the case. I have seen many people ignore an inadvertent bump or the sharp attitude of people on whom they were waiting. It must not be the person doing the bumping and growling but rather the person doing the reacting. This is an essential concept to grasp.

We all have triggers. We all have experiences that we resent. But people's resentments are different and based on their own personal experiences of life. They have their own personal triggers. That being true, if we could get to the root fear that has created our resentment in the first place and deal with it, it will no longer hold power over us, and we will not have to react negatively and violently to the event.

Our patron could have been pushed around by older boys when he was a child and was never able to do anything about it. He could have been a member of a large family in which the pecking order required physical confrontation, and the event reminded him unconsciously of that fact and triggered his old childhood reaction. Whatever the reason for the fear, the fear is something that has not been dealt with up to this time, so every time some event comes about that reminds the individual, consciously or unconsciously, of that frustration, there is a reaction. Often, the reaction is inappropriate and appears totally bizarre in terms of the current conditions, but it makes perfect sense in terms of the original trigger.

Think about your own resentments. What do the actions of others represent to you? Why do the things that irritate you do so? Why don't the same things that drive you to distraction irritate everyone? The answer is simply that not everyone has had the same experiences, so not everyone has the same resentments! They are all yours, and the reactions are all yours. Think about something that drives you to distraction. What really angers you? Pick something that is not very logical and that people wonder why you are so bothered by. It could be an accent, or a mode of dress, or having someone pass you on the highway. It could be a specific event or just a general condition in your life. Now look to see what it is about this resentment that you resent. What do you think is happening when the behavior or condition arises? Is it really what is taking place?

If someone passes you on the highway, what goes through your head? It could be that you've been challenged or that the other person thinks you're stupid because you drive slowly. It could be that you think the person is going too fast for conditions and that he or she is endangering others' lives. It might be that if you have to follow the law, you think everyone should. Or it may be that this is not one of your triggers and you just make note of the faster driver as part of your environment without any resentment at all. Now, how do you react? Do you want to catch up with the drivers that passed you and pass them, proving you're a faster driver than they are (competition)? Do you want to hem them in so that they can't drive so fast? Do you want to give them a verbal piece of your mind? That's all just a desire to seek revenge. Note that the persons in the other cars probably don't even notice that you're angry. They're just driving along, totally unaware of your existence, which for you may be another resentment, if you haven't dealt with your own feelings of insignificance.

Whatever the cause, there is a reaction to the event, and it has nothing to do with the other person. If you realize that and really accept it, then you are in a position to deal with it and no longer have other people's behavior in control of your experience of life.

Accept that the resentment has occurred but, rather than getting angry and seeking revenge, begin to look for the cause. When you find it, you will begin to realize that it is simply an issue in your own life and that your fear/anger is nothing more than a reaction to a lie. The other people and events of our lives are neutral events, and we simply frame them in accordance with our own fears to create the anger. The man who bumped into our hungry patron was not purposely trying to attack him; he simply bumped into him. A moment's reflection could have brought the patron to this realization before he went to the steps of anger and revenge, and he could have *chosen* to not let the event control his life. He could have simply noted that it was an example of that old program he had from childhood about pecking order, realized it had nothing to do with what just happened, and gone about his business. If he had, the results would have been far different. He could have had a pleasant meal and returned to the office, satisfied and calm, rather than hungry, physically ill, and irritated.

The choice is always at the point of resentment. At that moment, you can enter the revenge cycle that leads to more resentment and more revenge until your life is totally miserable. Alternately, the choice could be to stop, look at the truth of the experience, find the root cause of the resentment, and deal with that, rather than the immediate event. Then the resentment has a diminished capacity to control your life. This second cycle is a cycle of change, and it leads to a different experience of life, one that is more enjoyable and gentler than that of revenge.

Thus we may choose between revenge and change when resentments arise. What has this to do with being ethical? Is it unethical to respond to a threat? Is it unethical to confront those who do us wrong? What if I do not want to give up my resentments and feel justified in resenting them?

Be aware that any position you take is fine, as long as it is ethical. In actuality, it is okay if a position is unethical, *except that it does not work*. We have defined ethics and doing what works and have discovered that what works is to behave in a reciprocally loving way. As has been stated, the opposite of love is fear, and as such, any act of fear is an unethical act. Since resentment and the resulting revenge are acts of fear, they are by definition unethical. The whole reason we have discussed these concepts is that the fears that we harbor and do not deal with remain with us, altering our behavior and causing us to be unethical. To avoid this, it becomes a quid pro quo—*the more fears we can eliminate from our lives, the more ethical we naturally become*. Therefore, the only motivation left open to us is that of real love (not attachment), wishing well, and supporting the happiness and aliveness of others. Choosing an opportunity to change and to grow rather than one to seek revenge and remain the same decreases our fear and therefore increases our success in life.

COUNTERPOINT AND APPLICATION

This chapter assumes that all of our misery is self-imposed and that there is never a legitimate reason to attack another person or hold anyone else responsible for what happens to you. Yet there are people who do cause you injury and who do deserve to be punished (preferably by you) for the things they have done to you. Isn't that true?

That is not true at all. Understanding this chapter requires remembering the lessons of the past and the importance of changing paradigms. It is the norm in our society to blame others for our problems. If you do not feel that this is true, take a good look at the tendency of the majority of the population to depend on others—government, family, friends—to help them out when they are in trouble. Look at the litigious nature of our society, in which people sue at the drop of a hat and insist on more and more laws to protect them from the behavior of others, and the increasing expectation of outside forces to provide us with what we need. Look at the nature of your own tendencies to blame others; note how you really behave in this regard.

Be aware, however, that there is no one responsible for the content of your life but yourself, and that any attempt to blame someone else is merely a self-induced subterfuge, a rationalization designed to save you from facing the truth about your own culpability in the matter. We are the masters of our own fate, the authors of our own lives. We are the ones making the decisions, refusing to look at the truth of that which we fear and come to terms with our past experiences. Effectively, in blaming others and going through the TAR cycle, we are enlisting others as excuses for our own shortcomings, and whereas we have perfect free will to do so, it simply will not work. It will not cure our neuroses, it will not solve our problems, it will not keep us from the same fear and anxiety the next time a similar trigger shows up. It merely expresses our natural animal desire to protect ourselves when afraid by becoming aggressive and at some level violent. If that works for you, do it. And remember that it will be accompanied by a reciprocal behavior by others in your life. There is simply no free lunch, and in order to play, you have to pay.

In practical terms, if you start to notice how often you do not move in a given direction in your actions because you are afraid, you will find where your triggers are. You will begin to see what you are fearful of, and if you are able to pinpoint the exact fear, you will probably realize how ridiculous that fear really is. There is a saying that we fear what we most want to do. That being true, the way to get what we want is to do what we most fear. If you are involved in the creation of technology, what keeps you from your true creative potential other than the fears connected with really putting yourself into the project and going out on a limb? Why are you afraid to take risks when the only result is a greater sense of reality? If you are a user of technology, when do you not do what you feel is right? How often do you fail to refuse to participate in technological processes that you personally feel are unethical? How likely are you to later blame others for your own lack of courage and refusal to act in the face of fear? After all, all those perpetrators of the Holocaust were merely following orders, weren't they? Wasn't it unfair that they were punished when it was really their superiors' fault? Is there a fundamental difference, other than degree, between those last two questions and your own refusal to take personal responsibility?

EXERCISES

1. Think of a recent experience you have had with TAR. See if you can trace the movement from trigger to anger to revenge in this cycle. Now consider the trigger in light of what you have read. How much of your reaction was honestly due to the situation at the time and how much to conditions set down earlier in your life?

2. The violence we do to others is often so covert that we are not consciously aware that we are doing it. Consider times when your actions have "accidentally" affected others in a negative way. What was your real motivation for setting things up that way?
3. To avoid even unconscious revenge cycles, it is useful to deal with our anger in more productive ways. Next time you feel injured, rather than striking back or pretending that you do not care, try defusing it. One of the best ways is to do something of value for another human being without personal motivation. For a really powerful example of how this can work to deal with issues, pick a person whom you feel has injured you and do that person a favor of some sort. As you do, you will find your own "angst" subsiding and naturally become aware of what it is within you that has created the anger.
4. Sit down and write a nasty letter to someone with whom you are angry. Read it daily with great emotion until you begin to see how silly your words and thoughts really are. When you begin to laugh at your own absurd reaction, tear it up. Never mention what you are doing to the other party.
5. The next time friends are triggered by you and react, try ignoring the attack and help them (gently) to look at their motivations by sharing your own experiences with them. See how dramatically it changes the situation.

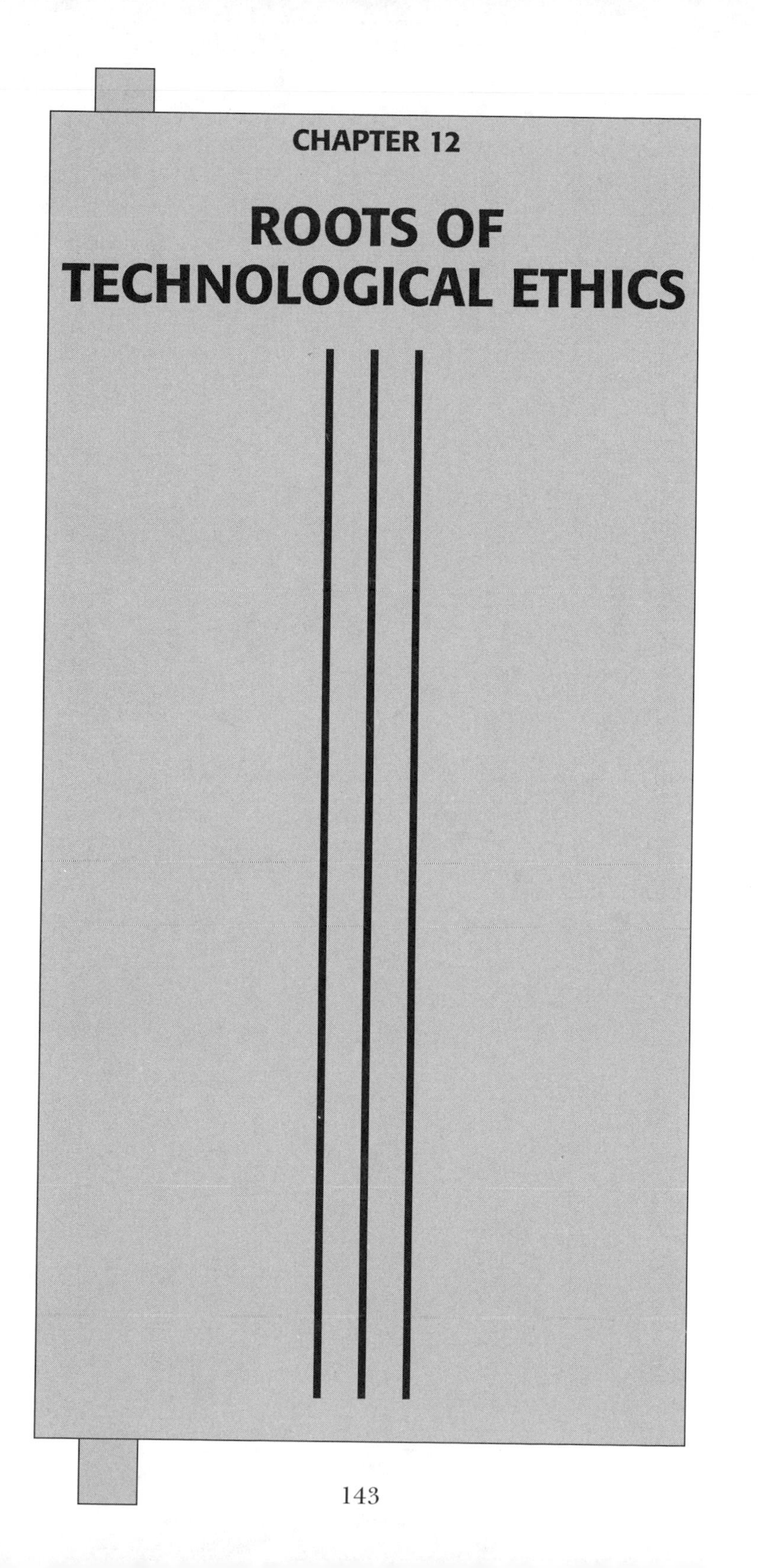

CHAPTER 12

ROOTS OF TECHNOLOGICAL ETHICS

INTRODUCTION

Be aware that technology itself is neutral. It has no more ethical content than does a tornado or a running river. Whatever the technological device or technological methodology, it is essentially a collection of interconnected objects that interact according to natural law and perform some predetermined set of tasks, according to the desires of the designers. We must concede, of course, that on occasion, technology performs unintended tasks and has unintended consequences, which is a large part of the ethical issues that we encounter in regard to technology; nonetheless the technology itself is essentially neutral. It is therefore not the *content* of technology that presents ethical problems but rather the *context* in which that technology operates.

THE ETHICAL ISSUES OF TECHNOLOGY

Technology results from (1) the observation of nature, which is a process of science, (2) the distillation of these observations down into principles of natural behavior, which we call scientific laws, and (3) the development of a wide array of applications of these laws or principles to do useful work. Technology is only a tool. It is a creation with a purpose and is systemic in nature, and it has no moral or ethical content in and of itself.

This last statement may seem strange if we consider some of the technology developed in the past, such as nuclear weapons designed for mass destruction, or the rack, a device for inflicting extreme pain on the victim. Yet as we shall see, in spite of the possible sinister nature of these technologies in use, the actual physical objects are free of ethical content.

The ethics of technology can be categorized as (1) the ethics of conceiving technological ideas, (2) the ethics of developing those ideas, (3) the ethics of creating the technology itself in physical form, and (4) the use of that technology for various purposes. In each of these phases of the creation of a technological event, there are ethical issues to be determined in how to go about the process. These decisions are influenced by both positive and negative factors, such as economics; security concerns; the ability to heal or kill, create or destroy, build up or tear down; ego; power; cultural impetus; religious beliefs; and pure resistance to change.

Technology itself is still neutral. It has no innate value except as we conceive and use it. Even a technology created for an expressly malevolent or benevolent purpose can be altered in its effect by how we choose to use it.

Technology will affect the entire system in expected, predictable ways as well as unexpected, unpredictable ways. Because of the systemic nature of the world in which we live, there is no way around that. We know that in solving a problem we may create others. Actually, in creating solutions, we create the necessity and opportunity to have other adjusting changes take place in the system. The true ethical issue lies in how we view the combination of what we choose to call positive and negative consequences and what this leads us to actually do. We must also be constantly willing to alter our view, our paradigm, in the face of new information. Nothing remains the same. Nothing will stay as it is, and fighting that principle is a losing battle. Growth is the only evidence of life, whether that growth/change is slow and evolutionary in nature or rapid and revolutionary in nature. (The main difference between rust and a stick of dynamite exploding is the speed at which the oxidation takes place). In any event, the change will occur.

In this chapter, the nature of this relationship between the creation and use of technology and the ethics of that creation and use will be explored.

THE SCIENTIFIC/TECHNOLOGICAL PROCESS AND ETHICS: THE PURITY OF SCIENCE

Technology is neutral. The very process of science is one of discovery. Its purpose is to investigate the nature of the world and to determine an increasingly complete and thorough picture of the mechanisms under which the system operates. The very concept of a system in operation is both philosophical and scientific, depending in the latter case on empirical evidence to support hypotheses concerning what happens and how events reflect natural laws. Pure science is called pure science simply because it has no motivation other than discovery. It does not pass judgment on the value of a fact or concept, nor does it directly concern itself with the application of that concept, except as such an application could lead on to further revelations about the "nature of nature." This concentration on pure knowledge is one of the elements that set science apart from other fields of study.

By dispassionately separating themselves from the consequences of their discoveries, scientists are able to philosophically and intellectually concentrate on the content of their chosen fields and restrict their studies to the nature of the phenomena they encounter rather than the possible uses to which that information may be put. Even when the motivations of the scientists are to better the world, they generally see their efforts as bearing fruit only by discovering objective truths rather than directing their efforts in one direction or another. The fact that a failed experiment in science is as fruitful an endeavor as a successful one, since the elimination of incorrect beliefs is as valuable as the discovery of correct ones in the pursuit of truth, is proof of this principle. It is exactly in this separation from consequences that the problem lies, because it tends to deny the existence of social, cultural, and environmental consequences to that work through the use of the discoveries by others.

Be aware that to some degree, this separation is an ideal more than a reality. The objectivity of the scientist is not pure. These are human beings, and no matter how dedicated they may be to the principles of objectivity, their personal desires and goals will affect their work. As Heisenberg pointed out, the very act of observation affects the results of those observations. What methods of observation are chosen and what variables are studied are the choices of the observer, and the conclusions drawn may only be unambiguous in the mind of the observer. Such behavior is more a sign of the humanity of the scientist involved than it is of poor scientific work. Bias is simply not easily removed from the research process.

Certainly, there are more reasons for studying science than mere curiosity. Particularly in today's world, much scientific research is funded by government and private, for-profit corporations. Thus, there is an ulterior motive, whether social betterment or economic wealth, driving the search for new information. The purpose is to find applications for the new information discovered; in terms of funding, this is often, far too often perhaps, the motivation for the search to begin with. When this occurs, the purity of science suffers. Because of the high cost of carrying out scientific research, there is an accommodation that takes place between the scientists who passionately pursue truth and the governmental or business interests who seek to profit from those discoveries.

There is nothing inherently wrong with a symbiotic relationship of this type in which each party is able to glean from the relationship the benefit that it seeks. This is in fact the essence of cooperation as opposed to exploitation, in which each party supplies the other with what it wants, and thus both parties benefit. The scientist is given the opportunity and means to pursue his or her own intellectual interests, and the funding party is given the benefits of any discoveries that may lead to new devices and technologies. It is a natural, systemic process in which each element supports the other. That is not the point here. The point is that because of this relationship, *scientists cannot separate themselves from the ethical issues that surround the application of their discoveries,* any more than technologists can separate themselves from the use to which their technologies are put. So much for the myth of scientific purity, at least in the modern world. Unless an individual is working alone in some basement laboratory or funded by some organization or individual who is not interested in profiting from the research in some way, scientists have culpability in the results of their behavior. In saying this, I run the risk of incurring the wrath of some of my scientist colleagues. Yet it is true nonetheless, and I am very happy to find that most of the physicists, biologists, chemists, and other scientists whom I encounter recognize that they are not exempt from moral consequences in their fields. At the base of their efforts is still an unending curiosity about the nature of things, of how this universe in which we exist and of which we are a part actually functions and according to what rules or laws.

To carry out scientific research in the modern world, however, the researcher must recognize and admit that every action taken has practical, real world consequences, and that this is as true of the discovery of new truths about nature as it is of any other human endeavor. Scientists are not responsible for the uses to which their discoveries are put, and at the same time, they must be aware that the unscrupulous, the economically motivated, and the politically motivated will undoubtedly attempt to take this new information and use it to their advantage. Paradox and dilemma arise for the discoverer of new natural laws because of this very fact. They are driven by their desire to know and their curiosity about the way things work. They are also aware, if they allow themselves to be, of the fact that the practical application of their knowledge may be detrimental to others. At this point, when we shift from the discovery of truths to the application of those truths, we enter the realm of the technologist.

TECHNOLOGY AND SYSTEMS IN ETHICS

When we turn to technology as a field of endeavor, we find a similar framework for ethical reference. Technology is the application of scientific principles. It is the use of our understanding of how the world is put together to improve our lives and our well-being. In truth, its purpose is to increase our peace of mind through the development and use of technological devices. It is, as are all other aspects of human activity, a method of improving our chances for survival. What needs to be remembered is that the difference between human beings and other creatures is that humans survive by the conscious manipulation of their environment to a much greater extent than other animals. Like all animals, we are armed with instinct that allows us to know basic survival patterns, but included in our repertoire of instinctual information is the ability to create artificial technological devices, that is, to *technologize*. It is our chief instinctual weapon for survival. This ability to technologize requires an understanding of our environment, and that is supplied by science in its search for an understanding of the way the system works. Thus through an application of science, we are able to survive.

Because of the systemic nature of the world (and apparently all of that which is physical in the universe), when we apply science to the development of artifacts (artificially constructed devices and objects that are different from the way the components appear in nature), we use those artifacts (technology) to provide us with what we need for survival; a ripple effect spreads out through the whole system. This has been a recurring theme throughout this text. Every element in the system both responds to and changes because of our manipulations. Simply put, our development of technology and the use of that technology have consequences for the entire fabric of the world. Every element in a system affects every other element in that system, and the more complex the system, the subtler those changes may at first appear. Yet persistence in the continued use or expansion of a new element in a system results in ever-increasing effects, and the overall response within the system is cumulative. Even apparently minor changes in systemic structure will result over time in sizable shifts in how a system functions and behaves. This appears to be unavoidable.

Thus, the development of agriculture using the swidden (or slash-and-burn) technique at first appears to do no more than clear a piece of land for planting. However, after three to five years, the land must be abandoned because the soil is worn out. Continuing this process indefinitely can have dire consequences. If we use the slash-and-burn technique as a means of providing food for the population, we find two results: The population increases because quality food is more readily available, and more and more land becomes infertile. In fact, as the process continues and the population grows (a systemic response to the new technology), we use up the land's fertility at an ever-increasing rate (another response to the new technology). Eventually, unless we are wise enough to allow the land time to recover, we run out of fertile land. With small populations, this is often not a major problem. If you are a member of a traditional society in the Amazon Basin, in which gardening is all that is required to provide the necessary cultivated foodstuffs for the population, you simply move over to a new piece of land or move to another part of the forest. Years later, the entire tribal group may return to the now renewed site to again plant gardens. With larger populations, however, we can ruin an environment with this approach. Larger populations are not nearly so mobile. They tend to create sedentary lifestyles supported by more or less permanent infrastructures, such as buildings and towns (more technology), that make it difficult to simply pick up and move. Larger populations also result in less available land for migration. This being the case, early civilizations, when they found themselves faced with dwindling lands to clear and increasing populations, simply responded with more technology to solve this new problem. They shifted from the slash-and-burn method to other forms of agriculture, initially what is known as horticulture, or small, permanent plots of cultivated land that is fertilized and farmed on a permanent basis. Later the shift was made to extensive agriculture, which is what we see in modern systems, where mono-cropping and cultivation of huge tracts of land is the rule. In each case, a solution was found, which brought with it both increased social welfare and more problems to be solved. This pattern continues. If we look at the world's population, its growth rate, and the rate at which food supplies are growing, we find that we are once again approaching a point of crisis. It may well be that just as the twentieth century was the age of petroleum wars, the twenty-first century will be the age of food wars and wars over rights to limited usable water supplies. If so, it will mean nothing more than one more adjustment to the problems inherent in our own successful manipulation of natural law.

It's a simple example, but it illustrates the principle well. Every new technology creates changes in the system, and whereas some of those changes are desirable (such as the increase in food supply), others are negative (as in the wearing out of the land). It does not take a tremendous amount of thought to find that this is true of other technology as well. Improved methods of fishing deplete the lakes and seas of the world. The continuing improvement in transportation systems in terms of speed and convenience results in increased accidental deaths, the rise of the drunk driver as a menace to be faced by society, and various forms of environmental pollution, not to mention cultural and social disruption on a grand scale. Our ability to communicate at great distances through technologies of telephony, radio, television, and computer networks both increases our flexibility and productivity and isolates us and increases expectations on the part of businesses and society concerning our productivity. It also increases our vulnerability to unexpected change in the complex technological system upon which we are dependent. All technology has positive and negative consequences, from the point of view of the user, simply because the system will adjust itself; of course, if we do not like the adjustment the system makes to our behavior, we call it bad.

As a result of all this action-adjustment-new action, we have an ever-changing world in which we are constantly juggling our responses. The question is (and it is very much an ethical one), Does it work? Apparently, the answer so far is yes, judging from the fact that we have been using technology as a successful means of providing us with our survival needs for several million years now. If there are negative effects as a consequence of using technology, we simply adjust, as does everything else, either by a modification of our use of the technology or by development of other technologies to deal with the problems.

In addition, many of the negative aspects of technology are handled by the manner in which we accept a technology when it is new. As mentioned earlier, humans have a natural tendency to fear that which is new. It is a protective mechanism. Instinctively, we fear the unknown and, through experience, develop wisdom in dealing with new situations and concepts that results in caution. This type of fear can be helpful if it does not predominate in our lives. It gives us breathing room when new technologies arise and an opportunity to evaluate and consider the consequences of that new technology in our lives. Within a free society, the introduction of any new product, device, or technological methodology tends to be slow at first. This can be seen clearly in observing the introduction of some new product on the market. Initially, those who are willing to take a chance on something new, people referred to as innovators, try it; these people are generally younger and more adventuresome or have specialized knowledge concerning the new product. The majority of the population will take a wait-and-see attitude, relying on the innovators to determine through use the value of the product for them. If a product is a good and useful one, one that increases the social and economic welfare of the population, the innovators continue to buy and use the product, and it becomes familiar to the general population. This familiarity reduces the degree to which the product is an unknown, and a second, larger segment of society, known as the early majority, will begin trying it. Eventually, the product becomes so commonplace that if it is truly useful, its use becomes generalized over the entire population for which it has relevance, and it becomes a standard. Marketers have been aware of and have been using this principle for many years. It is a homeostatic process by which growing familiarity results in growing acceptance.

The problem is that the level of complexity of the world system and the degree to which we can now manipulate natural laws to produce what we wish has reached a point at which mere homeostasis may not be enough. Change takes place too rapidly. The effects of any given major technological innovation on the cultural system is so rapid and so great that we find ourselves faced with serious problems before we have had time to analyze sufficiently the consequences of our actions. Additionally, the difference between short-term and long-term effects is such that what may appear benign in the initial stages of use and generalization may have hidden consequences that only appear after years or centuries of use. Consider the relationship between the ozone layer and the use of chlorofluorocarbons (CFCs) as a propellant in spray cans. It was only after years of use of the product that the relationship was found, and there is still some question as to whether the ozone deterioration is a naturally cyclical phenomenon or strictly the result of the use of chlorofluorocarbons. If CFCs can have so major an impact on the environment, it would be advisable to know it ahead of time, not after years of use.

In the short run, the dangers from CFCs or the emissions of automobiles and factories or the runoff from chemical fertilizers or other artificially created substances may be miniscule. It is the cumulative effect on a system trying to adjust to and absorb the newly introduced conditions that causes the problems. This does not mean that the use of CFCs, automobiles, chemical fertilizers, or plastics is unethical, merely that the excessive use of these items is unethical in that it simply does not work.

It would appear that we have a tremendous ability to produce technology and that this ability is natural and inherent as a survival pattern. It would also appear that because of the highly complex nature of our modern world and the rapid manner in which new technologies can be disseminated through human society, technology has become dangerous to the very survival it was designed to support. This sounds like an insolvable paradox. In the very nature of our success lies our folly.

THE ETHICS OF RATIONAL TECHNOLOGY

Technology is a tool of society. It is a cultural element as much as family, political structure, economics, or religion. It is an integral part of the fabric of what it is to be human and live in a human community, a necessary element without which who and what we are would be far different. The issue is not whether technology is ethical or unethical; to approach the question in that manner is fruitless. The question that needs to be answered is, How do we develop and use technology in an ethical manner, in a manner that works for us? This is a very different issue altogether.

This theme of how to develop and use technology in an ethical manner is by no means a new issue. Each time a major technological shift has occurred it has caused disruption to our lifestyle and culture. This is natural. We must remember that part of the process of culture is to transform who and what we are, and technology is certainly a large part of that process. It must also be remembered, however, that the purpose of culture is also to reproduce itself, passing on the successful methodology and social interactions that lend themselves to our survival as a group and as individuals. We are a system, and we operate within that system according to rules that tend toward maximizing efficiency and success. It is in this reproductive process that we find the ethical issues. It is part of the homeostatic tendency toward resistance to change that we find the roots of the reluctance to take on new ways.

When faced with new technology, particularly technology that is obviously and rapidly changing our way of life, the cry of those afraid of the change will be heard. New ideas may be seen as the "work of the devil" or as a sign of our doom. They may be seen as a method of creating misery and unemployment for many. This is resistance to the new in the face of unknown effects. Caution is wise with sweeping technology, and our natural tendency toward caution keeps us from moving too quickly. However, it also tends to create technophobia among the most conservative members of the society. Hence we find survivalists involved in bunker mentality and preparing for the end of civilization or the breakdown of social order. In truth, this is a possibility with sweeping changes in technology. Periods of great change in society are often accompanied by social unrest and violent upheaval, as was experienced in Europe as it emerged to a more rational philosophy in the sixteenth and seventeenth centuries. Chaos occurs at the point of shift, where the old gives way to the new. It is not surprising to find people resisting those changes that will so thoroughly disrupt their ordered, structured lives, based on a lifetime of tradition. Whether it is the invention of the printing press or the emergence into the industrial age with its steam engines and great factories, violent reaction to rapid change is not unusual. Every such change will have its Luddites, destroying the new to preserve the old. Fear is a great motivator.

As part of this resistance, the ethics of the technology involved in the change is often scrutinized. "If God had meant us to fly, He would have given us wings!" is a prime example of this. At various times, the printing press, the steam engine, the automobile, the airplane, television, and computers have all been viewed by some as the work of the devil, and they have been vehemently fought against by those afraid of change. Yet it must be reiterated that it is not the technology that is at fault; the manner in which it is used creates the ethical problems. Rational technology means conceiving of technology rationally, designing it rationally, choosing technology rationally, and implementing it rationally.

The people of any society are the clients of engineers and technologists. They have an implied contract with those who develop technology in that the society supports the efforts of the engineers and gives them the freedom to produce technology in return for value and support of the population's well-being. Additionally, there is an agreement implied that the engineers will produce items that are useful, efficient, and appropriate. It is also implied that they will create new and different technologies that explore the possibilities of creation to be evaluated by the members of society, usually in the marketplace. This is an extremely effective way of handling technological development in a competitive market. When any part of that process goes awry, problems develop.

We normally use the market as a mechanism for evaluating the usefulness of technology. In the marketplace, the technology can be tested; if it is found valuable, it is allowed to exist by virtue of people being willing to buy it, use it, and continue using it. If it is not found to be useful or desirable, it is excluded by dollar votes being devoted to other items rather than the technology in question.

In theory, this is fine, but in practice, it is much more complicated. As with the rest of society, our economic system is highly complex, convoluted in its actions, and slow to adjust on its own to changes in need. There are very large economic units in all facets of commerce that tend to dominate, whether the field is computers, automobile production, or banking. There is good reason for these dominant players, mainly that we look to bigness for efficiency and low cost. Being able to muster vast economic resources means the ability to invest in the necessary experimentation and development that brings about much of the new technology we experience. In return, these dominant producers and marketers receive control and high profits. The trouble is that they too will be homeostatic in nature, at once seeking new product ideas and new technologies to continue their profits and simultaneously resisting shifts to revolutionary methodology that they do not control. To do otherwise would seem to be a poor economic decision. No firm or industry wishes to support its own competition.

In addition, government that understandably dictates how this money will be spent supplies much of the research funds available. Its desires and views may be quite different from those of the engineers and research personnel seeking to create new and useful technology, and this can be stifling. Government tends to limit creativity by only funding projects it finds in its best interest to produce. Yet is not the government representative of the people? Is it not the collective manifestation of our consciousness and desires? And as such, is it not appropriate for government to guide the development of technology in accordance with those desires?

The issue here is not whether or not the government of the United States truly reflects the desires of the population. That is a topic for a far different discussion. The issue here is how an engineer or other technologist is to react to a world in which the practitioner's personal beliefs are in opposition to his or her agreements, duties, and job requirements. If I am making weaponry, is it to defend my country or is it to conquer and control others? If I am producing nuclear power, do the benefits derived from relatively cheap electricity override the potential calamity of a nuclear accident? If I am experimenting on animals in order to produce lifesaving methodology for hospitals and doctors, does that justify the suffering of the animals? How do you make the decisions necessary in the face of such dilemmas? That is the central issue.

Be aware that the decision is an individual one, and must be an individual one, depending on the consciousness and understanding of the person who has to do the deciding. As much as we would like to have someone tell us what to do, we must face reality. All our decisions are our own. We alone are responsible for our decisions. We alone are the ones who must live with the consequences of those decisions. And we alone must decide how we will choose to interact with this world.

Earlier I pointed out that the members of society are the clients of technologists. That creates an obligation on the part of the technologist to act in the best interest of those clients or to cease being a part of the agreement. It is a matter of reciprocity and a matter of realizing the connection among us all whereby each contributes to the well-being or pain of the rest. We cannot separate ourselves from those whom we affect, and that includes everyone. We cannot blithely deny the connections between our work and our actions and the lives of those around us. If we are to be responsible human beings (and remember that this means being able to respond), then we must understand that there is a contract among us all and that the way we carry out that contract—the way we fulfill our agreements with humanity—will determine the quality of life for all.

This does not mean that we must subordinate our own desires and needs to those of others. It means, rather, that we need to see the connection between our own well-being and that of others. It means it would behoove us to work for the mutual benefit of all by creating opportunities for freedom, wealth, health, peace of mind, and self-expression without demanding it, imposing it, or judging the choices others make regarding their own behavior. It is not our place to judge. It is not our place to say what others should or should not do. It is only our place to contribute to the process that which is positive, uplifting, and supportive of success. Such an example seldom goes unnoticed and always creates an opportunity for others to find a better way.

Choosing the word *client* rather than *customer* when referring to the contract between technologists and the people of the society was purposive. When you have customers, you are there to fulfill their desires and see to it that they get what they want. It is a matter of making things available to the public; they choose what they wish to have. With clients, on the other hand, the relationship is quite different. You are seen as the supplier of knowledge and tools. You are the expert, the one to whom the clients come to find what needs to be done, needs to be used, and needs to be instigated. You are not there to give them what they want. Quite the contrary, you are there to give them what they need, and they trust you to do just that. They do not always like what they hear and are not always happy with what you present to them, but they are there to find out what they do not already know and benefit from your expertise and creative ability. They may think that they will be happier by involving themselves in destructive processes. That may be what they want, but it is definitely not what they need if they are about the business of supporting their freedom, health, wealth, peace of mind, and self-expression.

How do you deal with these clients? What do you supply them? Do you create technology that slowly kills them or technology that improves their experience of life? What is your purpose in doing what you do, and how is all of this altered by the fact that we are usually serving multiple clients simultaneously? This is a reciprocal process as well, and whatever energy you put into your work, whether it be greed or anger or love or joy, it will be reflected in the product of your efforts. What do you offer your clients, and what do we as a culture offer those with whom we interact?

We are all engineers and technologists, in a sense; we all create the fabric of this world by thought, word, and deed. What goods do we demand? What concepts and directions in future technology do we support? What kind of world are we creating with our interaction with the earth, the environment, and our fellow human beings? The system acts and reacts, and we all contribute to that dynamism. When people look back on your life, what will they see and say about your contribution to the whole? How will they view your decisions and your successes and failures? What will they remember? Much more importantly, what will you say and remember?

COUNTERPOINT AND APPLICATION

If all technology results in negative as well as positive effects, should the ethical person use or create technology in the first place? Is the concept of the most good for the most people really bogus, since according to the text, a solution to a problem is not ethical unless everyone wins in the process? How can a practitioner ever expect to avoid creating misery in the world through his or her actions under these conditions?

Remember the neutrality of technology. It is merely a tool, and as such, its use is a separate issue from its creation. That being said, be aware that it is the motivation of the creator that partially decides the ethical content of the technology. Nobel had in mind benefiting humankind with the invention of dynamite. He had neither control over nor conscious awareness of the negative possibilities to which his technology would be put. His motivations were noble, yet some of the results were negative. In that sense, it is true that we always run the risk of our creations being used destructively or unethically. Whereas this is a concern, it should in no way be a deciding factor as to whether or not we participate in the creative process. Albert Einstein was a very great genius. He was one of the definitive scientists of the twentieth century, a benevolent pacifist, a deeply religious man, and a benefactor to all humanity. Yet because of the use to which his discovery of $E = mc^2$ was put and his encouragement of its development, he has been vilified by some in recent years as the pawn of the devil. This is a level of absurdity that is difficult to believe unless you consider the lengths to which people will go to foist personal responsibility off on someone else. Those who vilify him choose to blame him because of their own lack of moral character or their inability to admit simply that they are afraid. The creator and practitioner have no control over that aspect of technology.

What can be controlled is the participation in the creation or use of technology that is obviously destructive and an awareness that technology can have negative effects, leading the practitioners to take steps to minimize those negative effects as best they can. Producing an automobile is not an unethical act. Producing an automobile and cutting costs by reducing reasonable safety features such as seat belts and strong construction may very well be unethical, particularly if it is done without the knowledge of the purchaser. An interesting question is, If a poorly designed car were constructed and the public told of its inherent flaws up front, would selling it be unethical? In thinking about this, try to ignore attempts to save us from ourselves through the implementation of laws governing the characteristics of produced goods. Try thinking of it in terms of responsibility for personal action and the tendency of individuals to fail to do so.

EXERCISES

1. It is believed that all technology results in both positive and negative consequences. Examine the three technological inventions that follow and look for those negative and positive consequences. Was it the technology itself that caused the results or was responsible for how it was used?

a. Eli Whitney's cotton gin

b. Thomas Alva Edison's incandescent lightbulb

c. Nuclear fission

2. According to the text, society is the ultimate client of technology. Clients are not the same as customers in that they expect to hear what they need to hear and receive what they need to receive rather than hear and get only what they want. This being the case, what is the ethical behavior for a technologist when interacting with clients?
3. Consider the case of a scientist who discovers a new source of power, gained by gathering and concentrating the "free electrical energy in the air." What are his or her ethical obligations to society concerning his or her discovery?
4. Continuing the example from exercise 3, now consider the position of the technologist charged with putting the new technology to practical use. If this person is to behave as a rational technologist, how will he or she modify personal behavior in carrying out the assignment?
5. In the Counterpoint and Application section of this chapter, an interesting question was raised concerning the manufacture of goods and services with flaws or dangerous characteristics and laws governing such manufacture. As an example, consider the cases of automobiles and baby cribs. If we assume full disclosure of the dangers of each, are the laws governing their manufacture necessary from an ethical standpoint, and why or why not? Why do you suppose the laws were established?

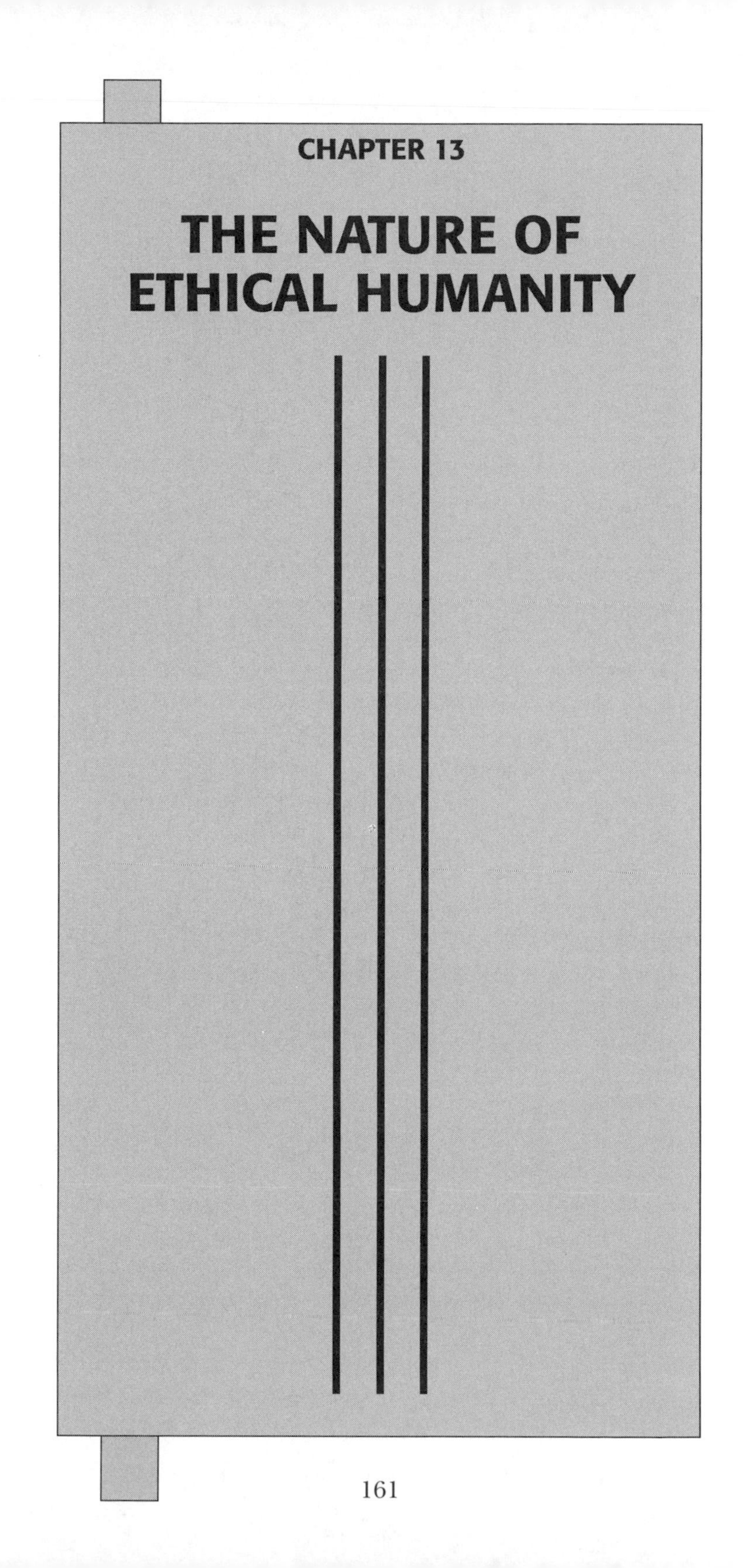

CHAPTER 13

THE NATURE OF ETHICAL HUMANITY

INTRODUCTION

How do you determine what an ethical member of society really is? From the former discussions, it would seem that an unending array of determining factors affects our behavior and that even after accepting the caveat to "do no harm," we are left with multiple dilemmas in our actions. How do you know if people are behaving ethically? As a guide, perhaps it is wise to look around and see what ethical humanity looks like. We can do this, indeed must do this, intuitively rather than intellectually. So many arguments and so many points of view lead only to confusion, and trying to set down hard fast rules as to what type of behavior is ethical is onerous at best. Yet we all recognize ethical behavior when we see it. We all know individuals whom we accept as ethical and whom we assume behave in ethical ways. How do we know this? How do we determine who is ethical and who is not?

COMMON TRAITS AMONG HIGHLY ETHICAL PEOPLE

We can recognize and even admire ethical behavior by observing the behavior of those around us and recognizing the patterns as either appropriate and successful or unworkable and unsuccessful. Some reflection on what we consider traits of the ethical person will go a long way in guiding us in our own behavior, with both technology and life in general. Consider for a moment a list of such *ethical traits*. It probably contains, at a minimum, the following twelve aspects of ethical behavior (each of which concerns a different aspect of development):

1. Integrity
2. Truthfulness
3. Courage
4. Focus
5. Decisiveness
6. Centeredness
7. Consideration
8. Creativity
9. Communicativeness
10. Acceptance
11. Awareness of perfection
12. Humility

Taken together, these traits seem to describe a personality that is unattainable for most of us. In fact, at first glance, they look like something out of a scouting handbook and may seem to describe an extremely dull person. Yet if we examine them in turn, we will see that the attainability of such a behavioral profile is well within our grasp and that the nature of an individual exemplifying these characteristics is anything but dull.

Integrity

Integrity is a matter of meaning what you say and saying what you mean. It is as simple as that. It is also essential for anyone who wishes to be in charge of his or her quality of life—to be truly responsible and experience life to the fullest. As we have discussed in earlier chapters, the world is essentially the way we believe it to be. Our words, our thoughts, and our deeds continually reinforce our view of the world because we are thinking, speaking, and acting in accordance with those beliefs. It is a self–fulfilling prophecy in which what we tell ourselves over and over is what we cause to come about. The interesting point here is that no matter how much someone tells us things are different from the way we believe them to be, until we truly and personally change our consciousness to embrace a different world, we experience our belief systems as they are.

It is as if there is a gatekeeper in our heads that stands guard and filters all of the information we receive and all of the things we tell ourselves. This gatekeeper compares the words of others and our own words with our unconscious belief systems and pigeonholes them as to whether they are or are not true. Interestingly enough, one of the main sources of the gatekeeper's knowledge of how we interact with our world is our own behavior. This is where the importance of integrity comes into play.

If I say I am going to raise my grade point average by a letter grade next term and I have resolved to do so, that does not mean it will happen. I can say I will get more done this week, clean the house, meet people for lunch, or be on time wherever I go, and none of it will come to pass if it is not internalized in the unconscious. The gatekeeper merely looks at my record, my experience in the past, and decides whether or not I mean it. How many times have I said, "I'll see you tomorrow," and not meant it? How often have I been late for a meeting? How often have I broken agreements made in haste or out of convenience to get someone off my back? The gatekeeper keeps track, and every time I break an agreement, I have neither said what I meant nor meant what I've said. Of course, there are times when we keep our agreements, but do these represent our behavior the majority of the time or only when it is truly important to us?

The gatekeeper does not delineate between important and unimportant agreements. To the gatekeeper they are equally meaningful. The gatekeeper judges our sincerity on the basis of our track record, and for most of us, that is none too good. This is unconscious for the most part. We speak casually. We speak out of habit. Yet we dramatically influence our lives in the process.

Notice that there are two parts to the definition offered for integrity. This is not by accident. Saying what we mean and meaning what we say are not the same thing at all. In the first case, saying what we mean, we are talking about speaking openly and honestly, without any attempt at subterfuge or attempts to soften our words at the expense of truth. To do otherwise is to lie, to dishonor the person to whom we are speaking, and to dishonor ourselves. We are lying because what we say is not truthful, and most of the time we know perfectly well that this is so. Telling someone that he or she looks good in a new dress when he or she does not may be the polite thing to do, but it is not the truth. We are dishonoring other persons by such an act because we hinder their desire to meet their goals (in this example, to look good). We assume that they can't handle the truth. How dare we make such an assumption? Think about it. Wouldn't you like to know the truth? If a new outfit looked really awful on you, wouldn't you like someone to tell you before you go out in public in it? Finally, we dishonor ourselves by limiting our own capacity for success. Remember that what goes around comes around. If we are merely polite or trying to avoid controversy, then others will do the same to us, not to mention that they will rapidly become aware that our words mean nothing and that we cannot be trusted. Thus, not saying what we mean, even for the sake of sparing feelings, simply does not work.

As for not meaning what we say, remember that the self-fulfilling prophecy that governs our lives stems from what we tell ourselves. The gatekeeper is listening. Often, stock phrases, expletives, and exaggerations are used in everyday conversation. Unfortunately, if they are repeated often, they become our true experience. This is a very subtle process. It happens unconsciously at a very deep level. Indeed, when brought to our attention, we are usually consciously aware that what we are saying is ridiculous and acknowledge that we do it for the purpose of emphasis or hyperbole. But the gatekeeper listens. If we say something often enough, for example, simple expressions such as "He really ticked me off" or "Life's just not fair," the gatekeeper will assume that that is the way we view life, and life will indeed become unfair. We will find it difficult to achieve equity in our lives because we will begin to unconsciously sabotage ourselves whenever we approach success. As for being ticked off, I am personally aware of situations in which people have developed physical ticks and other physical ailments as a result of less savory phrases and have successfully conquered their maladies by simply changing what they say to a more positive phraseology. Words create thoughts, and thoughts create and become manifest in the physical world. This is not magic; it is simply a very powerful subconscious mind at work. This is the reason it is recommended that the "words of my mouth and meditations of my heart" be acceptable. It is not just a good idea or a religious ideal. It is excellent practical advice!

As observers, we can tell when persons are speaking from the heart and mean what they say. We can intuitively tell when we can trust others to do what they say they will do and to tell us the unvarnished truth about what is happening. In business, this is a highly sought after quality in employees and in those with whom we do business. In our personal lives, it is essential to forming and maintaining workable relationships at all levels. And it is equally true that we know when someone is not trustworthy and cannot be counted on.

There is, however, an added benefit to having integrity. It tends to make life much easier. Things just tend to work. If you practice consciously being aware of what you agree to and then following through no matter what, you find that you begin to limit your agreements to only those things that you truly intend to complete. You will also find that as you continue the process and your "gatekeeper" finds you more and more truthful, you will be able to keep those agreements more easily and with less effort than before. Your unconscious mind will understand that you mean what you say and will work all of its "magic" to see to it that you do indeed keep your agreements, even if you forget. This will become clear by way of the following illustration.

I am aware of a colleague of mine who is very serious about integrity and has worked very diligently to improve his willingness to do what he says he will do. On one occasion, he and I agreed to go to lunch. I chose the time and the place. In fact, going to lunch together was my idea. He had intended to have lunch on campus that day. When we arrived at the restaurant, we met a friend of his, who said, "Well, I see you're right on time as usual. Let's go eat."

My colleague had made a date for lunch several weeks earlier and had totally forgotten about it, yet he "arranged" to keep the agreement unconsciously. Could it just be coincidence? Of course it could. Yet if this sort of thing happens time and time again, as it has to him, then it becomes obvious that it is simply the unconscious mind seeing to it that since he meant what he said, events would take place as agreed, even in the face of very difficult circumstances.

Truthfulness

Truthfulness goes hand in hand with integrity. The subtlety that lies in this particular trait has to do with honesty rather than resolve. Truthfulness does not mean simply relaying the facts or giving monosyllabic responses to questioning. Truthfulness requires a willingness to go below the surface of situations, to bypass fear, and to look at the truth of what is going on in order to effectively respond.

Truthfulness requires clear seeing. It requires an awareness of all of the finer nuances of people's behavior and motivations as well as the true nature of the interactions that are taking place. It requires, in essence, very focused attention. You cannot operate truthfully and simultaneously be "unconscious" regarding your life.

Most people operate on automatic pilot for the majority of the time. We perform the same functions over and over in our lives, meeting the same people every day, greeting them in the same way and receiving the same response, dealing with the same problems that have the same solutions at work, in traffic, or at home. This is where much of the boredom of our lives comes from, and moving through life in an unconscious, passive mode, not really aware of what's happening causes it.

Have you ever driven to work and not remembered the trip? Have you ever been reminded of things you've said and had no idea you said them? Have you ever suddenly discovered that a relationship was entirely different from what you thought it had been or that a friend was involved in a lifestyle that you never knew about? When you see someone and say hello, do you really pay attention, or do you just go through the motions, more interested in your own personal dialogue than in interacting with the other person? If any of these events have happened to you, and probably they all have, you have been living unconsciously during those times.

To be truthful is to be genuine. Indeed, one of the challenges of life in today's world is discovering how to be genuine in a world of deceit and misrepresentation. To be honest is to be wholly present in your life. To be able to respond to events appropriately and therefore be responsible, it is necessary to be aware, and awareness is truth.

All of this takes practice and a willingness to give up your own misconceptions about the nature of reality. You must consciously become more observant of what is happening, whether it involves the feelings of tension as you walk into work or the sense that a friend is experiencing pain or despair. It requires consciously being willing to look beyond the obvious to the subtle. In the process, you will learn to still your own inner turmoil and become part of the larger world around you. In essence, you will begin to include the world in your own awareness and will be pleasantly surprised at what a beautiful, colorful, exotic world it is. Those who learn to recognize the truth need never be bored, no matter where they are. Any environment, any location is so rich with information that boredom ceases to be an option.

Surprisingly, this can have a calming effect on the individual rather than one of overstimulation. That may seem an odd statement. How can an enhanced awareness of one's surroundings, other people, and experiences not create heightened stimulation? Initially, the stimulation does rise, but as you pay more and more attention to the processes and events of your life, patterns appear that bring all of these myriad events into perspective. When this happens, you begin to become calm in a natural understanding of how perfectly the universe functions and can concentrate not so much on the events themselves but on what the events tell you about yourself and those with whom you come in contact.

Courage

Courage is one of the more paradoxical principles of ethics, in that the more it is exhibited, the less it is needed. Courage is defined as acting in the face of fear. Most people confuse courage with fearlessness, but these are very different processes. The word *courage* comes from the Old French word *cuer*, meaning heart. Courage is indeed of the heart, and only in your heart can you find the bravery to act even though you are afraid. Courage is not a negation of fear; it is an overcoming of fear. Fearlessness, on the other hand, is an absence of fear, which in many situations may be antithetical to survival, if not an indication of recklessness.

To be ethical, particularly in a world populated by people who do not exhibit integrity, is most certainly an act of courage. The nature of being ethical is such that it often requires you to act in a way that is contrary to the norms of the society or the customs of one's social group. Most people live in fear and give into that fear. As stated previously, acting out of fear is by definition an unethical act; it is the opposite of acting out of love. Yet the opportunity to take the easy path and conform is very strong, even if we are fully aware that our own values tell us we would be better served by some other course of action. Group pressure, threats from other sectors of the culture, and a predisposition to conform to the norms of the group all conspire to override our intuitive understanding of what works. It is a constant battle, at first, to defy the majority and follow our own principles and values.

It is also dangerous to do so; those around us will not like being reminded of their own shortcomings. Nonconformity could result in physical harm or suspension of group membership if the group feels sufficiently threatened. Thus, even the tenets of Maslow's hierarchy tell us to take care of the lower level needs of safety and belonging before venturing into the realm of intuitive expression of our own individual beliefs.

If everyone is accusing an individual unjustly and you see the injustice of it, it takes courage to act in the face of that fear and stand with the accused. If the rules of society are unjust but accepted, even to the point of law, it is difficult to refuse to go along with those rules. If the culture within a profession or a business firm contains a tacit acceptance of unethical activity, to stand on principle could cost an individual his or her job. Yet the ethical person does so and does so with courage.

Mahatma Gandhi found injustice in the treatment of the citizens of India under British rule and refused to accept it. He fought it at every turn, nonviolently and with courage, even though he was thrown into prison, ridiculed, and badly abused. Henry Thoreau likewise fought injustice in his community, preferring to go to jail rather than pay what he considered to be an unjust tax. Martin Luther King Jr. fought the injustice of segregation though he was continually threatened with bodily harm and death. In each case, these men showed courage in supporting their own values. They had too much integrity to do otherwise. They were too aware of the truth to ignore it. They acted in the face of fear.

These are extreme cases. Few of us rise to the level of courage necessary for such undertakings. However, we still face moral dilemmas in our lives. It requires courage to risk termination in a job because of an unwillingness to carry out unethical business practices. It requires still more courage to resign from a position because of the unethical nature of that position. It requires courage to remain ethical in our dealings with each other, particularly when it may mean a loss of a friendship or spouse. It requires courage because we are afraid of consequences; we do not realize how in control of our lives we really are.

Fortunately, this need not be an ongoing battle. As it happens, one of the paradoxes of life is that the more courage you exhibit, the less courage you need. Acting out of principle requires courage only if there is fear, and as we act in spite of our trepidation, we invariably discover that our fears have been for the most part groundless. Courage is miraculously self-extinguishing in the light of truth. This is because there is really nothing to fear.

To understand this, consider the fear of getting fired for ethical actions. Our minds conjure up all sorts of horrible scenarios and plays them for us. We could lose our jobs by speaking out or refusing an unethical request. We could lose our homes or not be able to feed our families. We could find ourselves out on the street with no prospect for further employment and no hope for the future! With all that at stake, would anyone in their right mind dare to act out of principle? Of course they would go along with the crowd and follow the corporate culture. After all, they are only following orders. They have no culpability in the act. They have no choice. It's the fault of their bosses. And all that is supposed to make it okay to cheat, to steal, to lie, and to hide the truth. At least, that is generally the type of rationalization that we hear, assuming the employee links the issue with behaving unethically.

All of that is fantasy until the act is performed. It must be remembered that we live in a reciprocal world, and though consequences exist for every act, we are the ones that label those consequences bad or good or see an event as negative or positive. Suppose you do stand on principle and refuse to perform some requested act because it is clearly unethical. The possible outcomes could be nothing or heightened respect from peers and supervisors. It could result in a sudden shift in the collective consciousness of the firm (someone has to stand up and mention that the emperor is not really wearing any clothes!), or it could mean condemnation, a reprimand, an unfavorable review, or termination of employment. These are all possible. Yet if we look at these possibilities, it becomes apparent that none of them are in and of themselves "bad." Even termination may be a positive. If your firm is behaving unethically, do you really want to be a part of it? Reciprocity will eventually catch up with the firm, as it does with individuals, and when the illegal or unethical acts come to light, do you want your name associated with them? If a firm cheats suppliers or customers or lies to employees, how will it treat you as an individual? In all probability, you would view working somewhere else as preferable to your current employer's behavior, and either resignation or termination may actually be a way to find more suitable employment with a firm of higher integrity.

The courage stems from not understanding the reciprocal nature of the situation. No one ever suffered for his or her ethical stands. People suffer for their lack of understanding of the way the universe operates. No one ever said that a person must be a crusader, though he or she may choose to be. No one ever said that it is necessary to sacrifice and suffer, to become a martyr on the altar of justice. The only suggestion made here and in most of the literature concerning ethics is that you need not personally involve yourself in the behavior. When individuals have the courage to act in the face of fear, they discover that all of the terrible things that they were afraid would happen do not or that the terrible things that they were afraid would happen do occur but that they are a passport to freedom rather than something to be feared. Have you ever heard people say that being fired or leaving a job turned out to be the best thing that ever happened to them? Have you ever known someone who flunked out of a college or ran afoul of the law and reported later that he or she was glad that it happened because it straightened him or her out? Seemingly negative events are often positive ones in the light of hindsight. It is our fears that trap us and force us to have courage.

As you exhibit courage more and more, it becomes easier each time to take a stand and face the consequences of decisions. Decisions become more and more considered rather than haphazard, and the results are seen more and more in light of the truth—which is that motivations and our acts together result in our experience of life. In time, the temptation to behave in a manner that is not considerate of everyone, that is not win-win, goes away because it is only the fear that tempts; there is a realization that there is nothing to fear. With that realization, the courage to act is no longer needed, because the act becomes a natural expression of who we are and we are willing to express that nature.

Focus

This may seem to be an odd choice for a list of characteristics of ethical people, but if you reflect for a moment, you will notice that focus does indeed seem to be one of the expressed traits of the people in your life who appear to be ethical human beings. They always seem to know what they are doing and where they are going. Ethical people seem to seldom founder in indecision or not have a direction. To use a current phrase, they seem to keep a firm hold on their *moral compass* and not stray.

This is a practical trait as well. Focus allows a person to concentrate on what is important. People who accomplish a great deal seem to have purpose and seem to be on purpose most of the time. Additionally, they usually seem to do this effortlessly, without much conscious thought and with little resistance. This is a very logical result of living ethically.

If you recall our definition of a system, you will remember that all systems have a purpose and that the behavior of the system is designed to obtain and maintain whatever purpose the system's goal represents. This is as true of human beings as it is of any other system. We all have purpose. In fact, we have several. At the basic level, we have the fourfold purpose of our natural being, that is, to survive, thrive, reproduce, and transcend what we are. Even evolutionary theory tells us that without these four prime directives in our lives, we cannot exist as successful organisms. If we have no desire to survive, we become fearless and foolishly carry out activities that are dangerous or lead to our demise. If we do not seek to thrive, there is no growth, either personally or for the species, and that is a condition of stagnation that puts us at a disadvantage in a world in which only the most fit survive. (Be aware that this statement in no way implies the necessity of violence for survival of one at the expense of others. There will be more to say on that later.) If we do not reproduce, then our genetic makeup is not passed on to the next generation, and there is no continuance of the species, which is simply another form of nonsurvival. Finally, if we do not transcend who we are collectively and individually, we will express no growth as a species and will not be responding to changes in the larger systems of which we are a part. And systems theory shows that systems that do not adjust will die out.

Our purpose does not stop there. We have individual purposes based on our abilities, our talents, our inclinations, and the fact that we live in a cultural system that is about the process of surviving, thriving, and reproducing, as well as transcending. There is a tremendous amount of freedom available to us as to what we can do as part of the larger system and what we can do for our own individual growth and development. We have many forms of employment available to us. There are many forms of self-expression that we can choose, and we develop our lives around a set of behaviors and beliefs that support our well-being and contribute to our four prime directives. How we choose those behaviors and beliefs is a combination of individual talents, upbringing, cultural conditioning, and personal experiences.

This is the way we define ourselves. It is the way we discover who and what we are all about. As we develop over time, we discover what we enjoy and what we see as the truth of life and act in accordance with it. We find our purpose in life. We find the manner in which we wish to contribute to the community and to our own self-expression. All this happens naturally and gives our lives focus. It happens unless it is thwarted by our indoctrination and the belief systems of those who would influence us, good or bad, to their own way of thinking and unless events in our lives create states of fear that turn us from realizing the highest expression of what we are. This brings us back to the one element that can keep us from following our purpose and being focused: fear.

As we are raised, our parents and other influential people in our lives strive to give us the widest possible range of information about how things work and how best to avoid the pitfalls of life. All cultures do this. It is the process of reproducing the culture itself by teaching its members what works for their survival and what does not. Unfortunately, some of the training inherent in "upbringing" is based on the personal experiences and fears of those who are teaching us. If those who are our teachers are indoctrinated in a specific way or have developed fears from their own personal life experiences, then these doctrines and fears become inculcated in our own minds. Early in our lives, when we are most vulnerable and most trusting, this is a very potent force in our development. But what if the doctrines of our elders are at odds with our own natural inclinations? What if we are imbued with the same sense of fear as our parents or teachers, whether the fears are rational or not?

Consider an example. Suppose you are a young female and your parents are divorced (not a farfetched idea in today's society). Suppose that as a result of the trauma connected with the divorce, your mother, with whom you are living, has indoctrinated you to believe that all men are liars and all men cheat on their wives. This is irrational and illogical as well as simply not true (no matter what you may believe), yet consciously or unconsciously you believe it. Suppose that simultaneously your mother passes on the principles of her own upbringing (indoctrination) and instills in you a belief that you should find yourself a good and stable man—one who is honest and will provide a good living for his family—and that you should marry him and remain with him always. Immediately there is a paradox that simply cannot be resolved unless one of the two beliefs is deleted or modified. How does this affect your behavior? At a minimum, you will find a husband that you believe meets the requisite criteria for success and then proceed to destroy the marriage with fears of infidelity and dishonesty. At worst, you will live alone and unhappy, unable to establish any stable relationships or find an iconic man to fit your criteria for acceptability. None of this is probably in keeping with your natural purpose.

Consider a second example, in which a male child has an influential father who is a ruthless businessman, always ready to take advantage of another person, and pushes the definition of legality to the limit. Suppose this father has a work ethic that says, "Get them before they get you! It's a dog-eat-dog world out there!" Accept it as a given that this is antithetical to the child's natural value system. Watch a small child and you will see nothing but expressions of love, unless the child is directly threatened. We do not start out with manifold fears of the world and the people in it. We are taught to believe that way. When a child such as the one in this example grows to adulthood and enters the business world, the tendency will be to take on the attitudes of his father, unless he can free himself from the indoctrination enough to act on his own. The result is reciprocal misery, a constant searching for a better way to conduct one's business and personal life, living in a state of constant fear and aggression, and never being quite sure what works, even with all the available information. Such children have been trained. They've been taught how to run a business. They've been schooled in the finer points of devious business practices, yet they are never quite sure if what they are doing is good enough, effective enough, or going to work. They are never sure which direction to take. Essentially, they have no workable focus.

Persons who understand the way the universe operates, on the other hand, have no such problem. It is clear that they have a purpose in life, that it is their own personal purpose, and that that purpose is an expression of the individual self. They see their road as clear and their choices as simple. Every act is an act toward their purpose, and it is an act of love, both for themselves and for those around them. With a true understanding of purpose, it is simple to keep "on track" once a direction is taken, and it is simple to change direction if it becomes obvious that the path no longer leads to the fulfillment of one's purpose.

This focus is not a narrow stricture but rather an expanding field, in which all possibilities are acceptable as long as they lead toward completion of the purpose. It is freeing to not worry about whether what is being done is right or wrong, acceptable or unacceptable, good or evil. All acts of love that lead toward the purpose are okay, and it is not necessary to become locked into any single mode of behavior. At one time it may be appropriate to have a full-time job and a stable career. At another moment, it may be perfect to be a homeless wanderer. A college education may be the perfect step in the process today and the last thing that should be done at another time. The method ceases to be central to the actions of the individual. The focus is on achieving the individual's purpose.

Be aware that if it truly leads toward your purpose, your action will always be an act of love. I am not advocating any means as okay to achieve some goal. If you are truly looking at your highest expression of Self, then you will see it as an act of love for all involved and behave accordingly.

Consider your career choices. How were they attained? How did you come to the conclusion that one field and not another is where you should be? If it is based on anything other than the natural joy you receive from doing it, you are in the wrong field. People have a tendency to do things backward. They feel that they must do something to get something in order to be something. People believe they need to go to college to get a degree to be educated. They think that it is necessary to write fiction to get published to be a writer. This is backward. People begin by *being* something, which leads them to *do* something, which leads to *getting* something. A person *is* a writer, which causes the person to write naturally. If an individual writes long enough, it usually leads to a high degree of skill through practice, which leads to publication. Similarly, if a person *is* an engineer, that person will naturally carry out the activities of engineering, whether it be building dams in a creek or transforming old mechanical parts into some machine, which will further lead that person through natural curiosity to an education and then to a degree. The individual has been an engineer all along. These individuals have merely formalized the process and proven to the world what they are. There are many ways to express what we are; and we will express it. Of that we have little choice. That being the case, why not focus on what "turns you on"? Why not do in life what you love to do and not what someone says you should do or what someone says you should not do. Look for your purpose and your focus will follow, and that focus will lead you naturally down all the right roads to becoming what you are.

Decisiveness

Decisiveness follows naturally from the focus and realization of our own nature. Decision making is only difficult when we don't know what we should do. If we are ethical, that is, have integrity, are honest with ourselves, have recognized our purpose in life and understand that acts of love work positively in our lives and acts of fear work negatively in our lives, then making decisions is very simple. There is little or no confusion about which way to go. Every decision is based on its applicability to creating a win-win situation while achieving our purpose. It is as simple as that.

Indecisiveness can be thought of as the system's way of testing our realities. The word *temptation* comes from the Latin word for test. Indeed, it is tempting to consider alternative opportunities when they come up, whether it be a different sexual partner, another location in which to live, or a new job. How do we decide? We are *tempted* by the new opportunity. We test our worldview and our assumptions about how to live life and then make a decision. When we are focused, this test takes no time at all. If the new opportunity offers an enhanced method of achieving our purpose and is a win-win situation, then we take it. If it does not, we pass it up no matter how attractive or lucrative it may seem. It is only when we are not sure if a new opportunity is of greater value in the quest that it becomes difficult for us to make the decision.

This does not imply that all focused decisions are quickly made. If there is a clear delineation between what supports our purpose and what does not, then the decision is simple. However, if there is no clear-cut determination of whether the present course or an alternative course best serves our purpose, then the decision takes longer. That does not imply that we are indecisive.

Periodically, as people progress toward their purpose, they outgrow their present behavior. It is time for a change or, at a minimum, a reassessment of their definition of purpose. The test represented by choices is sometimes a method of achieving that reassessment. The person is forced to consider what the choices are and to more closely define what he or she is actually trying to achieve. A person may begin by believing that his or her purpose is to be a teacher and later find that actually it is to be of service to others, that the teaching is just a method for achieving that goal. If a better way of being of service comes along, the definition of purpose is more closely defined, and teaching itself becomes only one of many avenues to the goal. Now it is a question of what avenue best leads to the goal or whether both are equally valuable. Choice becomes easier, and *at the moment of choice, there is no vacillation*. There is no wondering what to do or if the right choice was made. All roads lead to the same goal, and we simply learn along the way. All roads have lessons for us, and we choose the ones that appear to get us there in the most effective manner.

Centeredness

The trait of being centered completes a triad of related characteristics among those who live ethically. By centered I am referring to the ability to remain on purpose, in balance, and cognitively aware of the greater picture, no matter what external events happen to be. Centered persons are more than merely focused in their actions and decisive in their actions. They are seldom taken by surprise and need little time to adjust to changing conditions.

Maintaining our center is a matter of balance. The balance stems from an understanding of the nature of things, a realization that external appearances do not necessarily reflect the actual conditions one is facing. In any case, if our behavior, words, and actions are genuine and in tune with our Self, then no conditions can cause harm. There is a calm that pervades the life of a centered person. There is an acceptance of life conditions as the consequence of past behavior and key to future development. There is a preoccupation with just experiencing what is going on and learning from it that precludes worry, fear, or any panic reactions. All that becomes inconsequential in the face of the opportunity to learn and become more in tune with who one really is.

This is not a state that is foreign to anyone but rather one that is difficult to maintain for most of humanity. We all experience being centered from time to time. Often it is the result of the circumstances themselves dictating the need for calmness and control that leads to the state. Or it could result from an inner feeling of just knowing that what is happening is appropriate and okay, no matter what it looks like. Being centered suggests acceptance of a situation without necessarily suggesting agreement. To accept what is happening and react to it does not mean you like it or condone it. It means that you are merely being realistic and honest with yourself.

In moments of stress or of great danger, there are those who have the capacity to remain calm and deal with issues while those around them are in a state of panic. There are those who just seem to know exactly what must be done and perform the necessary acts required to resolve issues or solve problems. Often this happens with an amazing degree of calm and clarity on the part of the individual. You have probably experienced this yourself, for example, when loved ones have become hysterical; as much as you would like to join them in their hysterics, you understand that someone has to remain calm and deal with things. You find yourself remaining calm and taking charge without even thinking about it.

Moving away from the extreme case, consider the more average case of settling down beside a stream or sitting on the side of a hill to just relax and be. If you've ever taken the time to daydream or to contemplate a peaceful scene in nature, you know what it is to be centered. There is a feeling of wholeness, of completion that comes over people under such conditions. There is a feeling of the present that eliminates thoughts of future or past. Just being where you are in the moment is sufficient, and you begin to experience it totally, observing, feeling, and watching. In such moments, you are centered.

Now consider those whom you know who never seem to be bothered by events and are simply not given to outbursts or fits of pique in their lives. These are not people who are simply in control of their emotions, keeping them in check and refusing to show their true feelings about things. They are genuinely and honestly calm and unperturbed by what happens in their lives. For such people, they are serenely pleased when life goes well. When it does not, they tend to be contemplative and accepting. In any case, they simply deal with life as it comes. Being around such people is simultaneously intriguing and calming. We tend to be drawn to them for their apparent peace of mind. We like being around them because they have a calming effect on us. In our hearts, we very much wish that we were like them in this one respect.

The reason that these individuals are serene is that they are living from their centers, from their hearts and not from their heads. They have successfully gone "out of their minds" in the sense that they are operating at a deeper level than others and, as such, experience everything in a deeper way. The meaning that their experience of life has for them is not the same as that of the average person, who busies himself or herself with solving the problems of day-to-day life and actively thinking ahead to head off the next crisis that may crop up. Yet these people are vitally involved in their lives and filled with energy that seems to come from a never-ending wellspring. Their energy seems to increase rather than diminish with effort, and they draw upon their experiences in a joyful celebration of life.

This is a rare talent and one that you will not often see in your life, but the degree to which you or anyone else can master the art of being centered is a measure of your honesty, your understanding of reality, and your acceptance of those around you. One thing is certain. When you come in contact with such people, you respond and are positively affected by the contact.

Consideration

Respect must be earned, but courtesy is everyone's due. So goes the saying. As with most old sayings, it is rooted in truth, and it is the truth of the statement that gives it such long life. Courtesy is a form of consideration, based not on knowledge of the individual but rather on a basic principle of life, that our actions should take others into considerations.

Apparently, if consideration of others is an element of the behavior of an ethical person, it must work. If so, then the concept of consideration must fit into our general understanding of what it is to be ethical. No matter what approach we use to determine ethical conduct, this turns out to be true.

Consideration benefits everyone. Systemic theory tells us that reciprocity is absolute, which would indicate that if I am considerate of others, they will in turn be considerate of me. It is a straight reciprocal behavior. Additionally, systemic theory states that systems are synergistic, that is, that the whole is greater than the sum of the parts. Those parts interact in such a way as to bring about a result that is greater than the added individual efforts. It is in the relationships among the elements of the system that the strength of the system arises and maintains systemic balance. Consideration of others is nothing more than a realization that we must take into account the other elements in the system if we are to be successful not only individually but also collectively. Since we are all connected and inexorably intertwined, our success as individuals is truly dependent on the success of others, which, of course, means that those others must be considered when we make decisions.

Ethical individuals are consciously or unconsciously aware of this important truth; thus, they do not make decisions exclusive of the welfare of those around them. This seems obvious, yet if we look at human behavior, we find selfishness, greed, avarice, and for some a strong desire to take advantage of every situation that leads to their own gain at the expense of others. Not everyone is wise enough to see the value of consideration.

If you give some thought to the concept, you will probably remember times in your life when being considerate has paid huge dividends and when not being considerate has cost you considerably. A close friend of mine often says, “Leave them smiling. You may have to argue with them later.” Admittedly spoken tongue in cheek, the saying is still a powerful reminder. I am aware of several instances in my life in which I have chosen to remain polite and calm in the face of considerable provocation and found later that the kindness was returned many times over.

Also, being considerate tends to create positive feelings about ourselves and by extension about the world in general. The world can be a gentle place, but only if we decide to behave in a gentle way. Aggressive behavior and opposing others leads only to further confrontation. This does not imply that we should accede to the wishes of every person with whom we interact. It is just that taking a moment to consider others’ points of view, to attempt an understanding of why they are behaving as they do, will go a long way toward removing the barriers that separate us all from each other. Consideration is a form of empathy, in which we recognize the worth and value of the people with whom we interact. It is an admission of connection and a method of understanding others and ourselves so that we can all move forward in our lives. This is a purely practical process. No one person can be successful without the success of others. It can be viewed as systemic reciprocity or karmic or the will of God or from any other point of view, but the truth is that as a species, we are bound to each other in a cooperative process. To deny that and consider only our own needs and desires is the road to disaster on every level.

Creativity

The observed creativity of the ethical individual comes from several sources. It is true that there are those who are far from ethical yet very creative. In fact, there is creativity inherent in the very process of being unethical, particularly if it is conscious behavior. Consider the case of master cheaters. They have a thousand ways to cheat on tests, assignments, and reports. They work diligently at discovering new methods to beat the system, corrupt the equitable processes of life, and achieve unearned success. In their minds, the success is earned, of course. Look how much trouble they have gone to in discovering the methods of cheating that they employ. Yet is this really success? They have been very creative, but the results are a lie, and the reciprocity of the system will take away whatever gain they have achieved many times over. The individual who falsely obtains a college degree, for instance, has nothing more than a piece of paper. It represents no skills, no abilities, and no capacity to produce, a fact that is readily proven in the workplace.

This is not the type of creativity to which I am referring. Creativity as used by the ethical person is expansive and supportive of self and of others. It is a creativity that stems from love and joy rather than from fear. It is a creativity that is effortless to find and use and brings renewal to the person doing the creating.

As human beings, we have little choice about being creative. It is in our nature. Just as the technology we use stems from our innate curiosity and the application of that curiosity to find new ways to achieve our goals, our curiosity causes us to be creative in all other facets of our lives as well. We have incredibly powerful minds, capable of great feats of mental and physical manipulation. The mind is a machine that is designed to be creative, and creative it will be, whether for our betterment or for our destruction. Whichever way we choose to go, either to act out of love or out of fear, the results and the fruits of our efforts will be obvious.

With the ethical individual, the manifestation of this creativity is beautiful to behold. It is as if there were two gardens side by side, one of which was approached in love and the other in fear. The garden conceived in love will be a beautiful sight no matter what kind of garden it is. The purpose may be ornamental as in a flower garden or practical as in a vegetable garden. It doesn't matter. In either case, what will be observed will be a beautiful creation. It will tend to be neat, organized, very productive, and a haven for all who enter. The creator of such a garden will naturally take the time to care for the plants there and for the soil so that there is a successful crop even under adverse conditions.

The garden conceived in fear will be quite different. The creativity that has gone into producing it will still be obvious, yet the results will be ugly and unproductive. The tendency will be to push the soil, to overburden it and plant too much in too little space. Attention will be given to production alone without regard for the aesthetics or harmony of nature. All this leads to disastrous results. Beauty is inherent in natural abundance. There is a practical reason for neatly laying out a garden and not overburdening the land with plants. It allows for the growth and development of the individual plants. Limiting what is asked of the soil is equally practical in that it allows the soil time to recover and to continue to develop its fertility. With the garden born of fear, the tendency is to add artificial fertilizer, to produce to the limits of the garden's capacity, and to remove everything that is not directly productive for the benefit of the creator. Such people work against themselves and their efforts are fruitless.

Consider a firm that develops a new product for the sole purpose of maximizing profit. The expression of greed is an expression of fear, since they are attempting gain at the expense of all else. They are not considering the people who will use the product. They are not considering the environment or the future consequences of their acts. They are simply looking to make as much money as they can and do not understand that that requires cooperation and support of other people's profit as well. The result is a product that may be valuable but that is produced using the cheapest raw materials by people paid the smallest wages that the firm can pay. Also, the promotion of the product is likely to involve questionable methods and suspect claims of benefit. For a short period of time, these very creative efforts will result in wealth. Yet quickly, as the wealth arises, it is destroyed by a market that will no longer believe false claims, by equipment that wears out from poor design and overwork, by employees who unionize or quit to escape the yoke of their poorly paid jobs, and by investors who quickly tire of the game.

In contrast, an equally creative person who sees a need for a product or service and develops such a product or service to benefit the public and make a profit in the process will find high levels of success. The well-made product tends to be valuable and cheaply but efficiently made. The purpose is to make a profit. That is true. But the method of making that profit is seen as a process of providing value for value. The public finds a valuable product available to it that fulfills its needs. The workers take pride in their work and are compensated fairly for their efforts, in no small part due to the creative methods developed by the entrepreneur, which allow him or her to pay higher wages. The investors receive high profits because they have invested in an enterprise supportive of the system, and the original creator has not only physical profits but also a level of contentment with the manner in which those profits were produced.

Part of an ethical life is an awareness of the needs of others and an enhanced creative capacity to find solutions to life's problems. It is a natural consequence of the mindset of doing what works. The better such individuals understand how the system is put together and the more they experience consciously the results of their decisions, the more they tend to look for opportunities to be of service and to use their talents in new and more productive ways. As counterintuitive as that may sound, it is nonetheless the truth. Ethical people are creative. They have no more choice in the matter than nonethical people. That is a part of being human. What makes it unique and what makes it so demonstrably noticeable is the manner in which they choose to be creative and the results of their efforts.

Communicativeness

Being communicative is another natural extension of ethical behavior. This is because communication is a natural process for humans. The difference between the communication skills of an ethical person and those of an unethical person is that for the ethical person, communication tends to be effortless and without conscious decision. I do not mean to imply that ethical persons do not think about what they say but rather that the process by which they choose to communicate is different.

The key to understanding this process lies in Maslow's concept of self-actualization. As indicated earlier, self-actualization is the process of expressing who you really are. In doing so, such individuals develop a high degree of integrity; they say what they mean and mean what they say. It is actually *the art of being genuine in a world of fear and deception,* and this is indeed an art as well as a skill. It is probably the ultimate example of pure creativity. This is what self-actualized integrity means. Under such circumstances, the thoughts of the person of such integrity naturally express love rather than fear, and the verbalization of those thoughts can only bring forth positive results. Without the fear, these individuals see no need to couch communication in polite terms to avoid offending others, both because they understand the truth of what they are saying and because giving offense is not in their consciousness. Thus it becomes a simple matter to speak from the heart.

We have all experienced this, whether we are aware of it or not. At times in our lives, we have spoken without first weighing our words and said things that surprised even ourselves. It is as if the thought came from somewhere else, and we were as astounded as our listeners at what amazing thoughts we uttered. An excellent way to see this process in action is to refer back to something you have written years ago. It will seem as though you are totally unaware of the concepts and are being exposed to them for the very first time, yet the words will have been written in your own hand and from your own mind.

At such times, we are given insight into our own internal clarity and how powerful words spoken from the heart can really be. The truth is that the more we become ethical, the more clearly we speak from the heart, and the less we allow fear to control what we say and how we say it. The communication becomes natural; it becomes spontaneous and more powerful to the benefit of all concerned.

Acceptance

Many philosophies speak of the importance of acceptance. Discovering how to be accepting and how to deal with the paradox of acceptance versus rational judgment can be confusing. The confusion that creates the paradox stems from a misunderstanding of what acceptance actually is.

We hear people say that we should accept others for who they are. We hear that it is important to be accepting and not judgmental. We are told that those religious and spiritual figures in the history of the world who were the paragons of the "virtue" we are seeking were all accepting beings. At the same time, we are urged to rationally judge the value of our behavior and the behavior of others. We are told to choose the good over the bad, the righteous over the evil. In this book it has been suggested that we will be far more successful, even on our own terms, if we choose what works over what does not work. Yet all of these ideas seem to suggest judgment. How can I be accepting and not judgmental?

This paradox, like all paradoxes, is only apparent. The truth lies in understanding what is meant by accepting and by being judgmental. *Acceptance does not require agreement;* it never has! No one has said that we should accept the behavior or beliefs of another as right. The only thing that is asked is that the behavior be accepted for what it is, the idea seen for what it is. Acceptance of the *truth* of an idea of behavior allows us to truly judge its worth. Without looking at things with a critical eye, there is no way to make choices among opposites. Acceptance, therefore, is a matter of looking at what is truly involved rather than accepting it out of hand without looking at all. The unexamined life is not worth living!

As for judgment, that is not the same as being judgmental. When we judge, we weigh all of the information about a situation, a person, or a concept and reach a conclusion (a judgment) based on that process. We accept it or reject it in our own lives based on our conclusions of how it fits our criteria. In matters of ethical consideration, we reach a conclusion as to whether or not it contributes to our success, that is, whether or not it works.

Being judgmental means passing judgment on whether or not the person carrying out the act is good or bad, righteous or evil, or right or wrong. Judgment is on the innate value of the individuals rather than their behavior or beliefs. Ethical persons are quite willing to accept behavior for what it is, judge whether or not they agree with it, and then separate the persons from their behavior. Ethical persons understand that an individual is not his or her behavior and that though we may feel that an individual's acts or beliefs are wrong, that does not make the individual bad or evil. There is no need for judgment in that sense, since the universe will reward an individual reciprocally, no matter what his or her behavior happens to be.

This may seem like a subtle difference in meaning, but an illustration may suffice to show that it really is not. There are many who could be chosen to illustrate the difference between acceptance and judgmental belief, but one of the best is the very real-world example of witches.

If we trace the derivation of the word *witch,* we find that it comes from the Middle English term *wicke*, meaning bad, which in turn comes from the Old English term for wizard. It is also related to the word *weird,* which comes from the Celtic word *weir,* meaning knowledge or wisdom. What we have by definition is an evil person who is knowledgeable and wise.

Of course, we all know about witches. They are those sinister old crones who ride broomsticks through the night and use spells and magic to wreak havoc on the rest of the population. They are evil through and through, worship and consort with the devil, and make huge cookies out of lost little children who run away from home. (Now that's what I call reciprocity!) At least that's what legend and early European Christianity would have us believe.

In truth, a witch is a person who uses incantations, natural herbs, and spiritual creativity to affect the physical world. Witches are very similar to shamans, priests, and medicine men in that they intercede on behalf of the rest of us through the use of specific knowledge not generally available to the public. Any anthropologist will tell you that this is true. The question is, how did our present perverted concept of a witch come about?

Since our society stems primarily from a traditional European cultural background, we need look no further than that background to find the roots of viewing witches as evil. It develops out of the fact that though most cultures have some form of witches in their past, the European witches (and warlocks) found themselves in direct opposition to an increasingly powerful Christian church, which was diligently attempting to convert the pagan populations of Europe to the "true faith." These pagan beliefs, which we collectively refer to as *Wicca,* had both a god and a goddess and a highly developed system of theology, belief, and ceremony that went back for thousands of years. The Druids were a class of priests for this religion, and they were very weird in the sense that they represented the store of knowledge of how nature works and how to manipulate it for the benefit of society. They were the judges, the teachers, the priests, and the practitioners of science of the Celtic culture. Obviously, they were evil in the eyes of the Christian church.

In addition to the Druids, the culture also had herbalists, spiritual and natural healers, midwives using natural cures, and other wise old women, who through their longevity had developed knowledge of how the world works and how to read natural signs. These old women (i.e., crones) were the witches of which we so often hear.

In the eyes of the early Church, anyone who did not agree with the tenets of Christianity was by definition lying and therefore worshipped the devil. Interestingly, this is literally true, since the word *devil* originally meant liar (or deceiver). To the early Christians, anyone who did not support the gospel was a liar and therefore a devil.

Be aware that the pagan religion predominated in non-Roman Europe, and virtually all of the traditional cultures of Europe followed its tenets. The Christian church had to portray the practitioners of paganism as evil in order to justify destroying them and replacing them with their own ecclesiastical structure. It was a practical move on the part of a religious system that firmly believed it had the key to understanding God and was therefore justified in doing away with false beliefs whenever and wherever it could. Thus to be a witch became synonymous with being evil.

This view still predominates in the folk culture today. Witches are the subject of legend, horror films, and books and remain the primary icon of Halloween, a Christian celebration, occurring the day before All Saints' Day, in which the dead are supposed to come back to life and terrorize the community until morning. Only by appeasing these goblins and witches with sweets and other treats can a household expect to survive.

Modern witches are everywhere. They can be found in virtually every European culture and in the United States from coast to coast. These are merely practitioners of a religion different from Christianity; they do not worship or consort with the devil (indeed, in the Wiccan religion there is no devil), and they do not go about the countryside wreaking havoc on the population. That is as untrue today as it was when first purported. Yet people fear them out of ignorance. People who are unfamiliar with the beliefs of Wicca still see witches as evil and dangerous. If they knew a person was a witch they might very well shun that person. In fact, to show how true this is, check your own perceptions of the fact that the author is writing this material. Have you begun to harbor a suspicion that I may be a warlock or that I may be Wiccan myself? As it turns out, I am not. My religious beliefs do not enter into this presentation at all. It is merely an illustration of how easy it is to make presumptions based on a limited number of observable facts and then judge people on the basis of that limited knowledge.

We can accept that witches exist and that some individuals are actually witches. We can also decide that we do not agree with their point of view and reject the teachings of Wicca as untrue. That does not mean that witches are evil, bad, or in league with the devil and should be condemned. Such a decision would be to judge their worth compared to others who agree with us. I can accept who they are without agreement and reject their philosophy without believing them to be evil. If I were to judge everyone who did not agree with me as *personally* damnable rather than damning their behavior, I would find myself in a world of damned people, which would most certainly speak to my own self-imposed damnation.

To further illustrate how incorrect it may be to judge true witches as evil, consider some of the beliefs of the Wiccan religion. They include a belief in the sanctity of all life. They believe in the reciprocal nature of the world and that what you put in is what you get back. Traditional Wiccan justice requires compensation for transgressions rather than punishment. Wiccans believe that men and women are equally valuable and free in the society. They are also conservators of the environment and take individual responsibility for their experience in life. I'm hard pressed to find the evil in any of these beliefs. The judgment appears to be a matter of religion rather than ethics, of dogma rather than truth. Even that is okay, if we accept that some people see things that way. You may agree with them or disagree with them, but it does not make you good or bad any more than it does anyone else.

Awareness of Perfection

One of the shifts in consciousness that takes place as a person moves toward a more ethical existence, as the perfection of life becomes more and more apparent, is a growing sense of beauty in the world. We live in a very complex and very responsive world, a system that contains so many elements that it is virtually impossible to see all of the connections that exist. Everything is connected to everything else. Because of the systemic nature of our world, there is no such thing as an act or even a thought that does not have an effect on the rest of the world. We are truly bound to each other inexorably. That is part of the reason for the importance of such qualities as acceptance, communicativeness, consideration, and integrity. Who and what we are affects everyone and everything.

That can sound very ominous, as if we must watch our every move to be sure that we do no harm to others. And some people, when they begin to realize the degree of the connectivity, believe this to be so. Yet the ethical personality does not see it as ominous at all. This is because ethical persons realize their own place in the scheme of things and that they individually cannot in any way create either pain or joy for another, only contribute to it. Individuals create their own worlds. Yet the universe is so perfectly structured that as we each go about the process of creating our own experience of life, we "magically" find ourselves in contact with others who will support that experience. Thus if we have a positive view of the world, we attract positive people and events into our lives. If we view the world as a terrible place, we attract like-minded individuals. It is not magic at all. It is merely the machinations of an immensely complex system adjusting to internal change. We call it magic or weird or divine or fate because the system is too complicated for us to see how we have brought things about.

The ethical person begins to see the connections not as discrete pieces but as overall effects of discrete actions, and the realization is comforting. It is a matter of noticing how perfectly the events of our lives reflect what we have put into the process. It is a matter of becoming aware of how balanced the process is moment to moment. Those who are doing what works know that they have nothing to fear as long as they continue to tailor their actions to what works.

Such individuals know that regardless of what circumstances may appear to be, they are perfect for the moment and that if a person merely waits a while, the perfection will become clear. Knowing that the moment is perfect, conscious persons do not ask why something is happening; rather, they ask how their behavior has created it. They do it without judging either the events or themselves, realizing that what we may call bad may ultimately be for our good and that what we call good may in fact lead to something we do not like. It ceases to matter, as long as the observer can benefit from the experience and use it as a guide to future actions in life. Once this is realized, the person moves to the twelfth and final quality of an ethical life, that of humility.

Humility

When a person begins to understand what a benevolent place we inhabit if we simply choose to do so, the state of awe that develops in that person is truly amazing. When the perfection and beauty of the world in which we live is realized for the first time, the awareness is staggering. If you have ever looked into the night sky and become aware of just how vast and marvelous this universe is, you have experienced this awareness. However it happens, when this awareness develops, simultaneously the importance of individuals is seen as everything and nothing. The need to struggle and strive for happiness is seen as unnecessary. The attainment of whatever is desired is seen to be a simple process, not hopeless or even difficult. All that is required is an acceptance of the perfection and a willingness to move toward it armed with the wisdom of experience.

This will create a state of awe. It renders you humble, and in your humility, you find a level of freedom that is not otherwise possible. All barriers dissolve though the contents of life remain the same. All need to understand shifts from rational problem solving to simple small steps. It is not necessary to do it all. It is no longer required that one develop detailed plans and schemes, taking into account all of the miniscule problems that may arise. All that is truly required for success is an agreement with Self to move forward in a way that works for everyone, a goal that is founded in a desire for the well-being of all concerned, and a strong feeling of love for the process of life itself. All else becomes a parade of passing events to be experienced and enjoyed. We find ourselves in charge by giving up control. We find ourselves in possession of what we want by not seeking it. We find ourselves with wealth, health, and peace of mind. We discover that we can have everything by supporting others in their quest for these qualities. We find ourselves.

COUNTERPOINT AND APPLICATION

About the only criticism that can be leveled at the material in this chapter, and it is a sizable one, is to point out that no one expresses all of the ethical characteristics all the time and that it seems rather pointless to even entertain the thought that they might. Be aware that whereas this may appear to be true, the fact remains that with the exception of psychotics, who are so buried in their fears that they can see no reality beyond themselves, virtually all of us express each of these characteristics at various times. The trick is to be consciously aware of when we do not and modify our behavior. Life is still an opportunity for learning, and as we offer ourselves the opportunity for new experiences and the results of new behaviors, we find that it takes less and less courage to behave ethically, because we become less and less afraid of the results. The more we embrace these behaviors, the more we experience them working to create peace of mind and all that goes with it, and they soon become more habitual. This is not to say that we will soon never behave out of fear. It is simply to say that the more we embrace these behaviors, the more content we will become. Fortunately, we have a lifetime to work on ourselves, and the adventure itself is so exciting and so fulfilling that I would not want to deny anyone the opportunity to be on that road.

EXERCISES

1. For each of the professions listed, determine how a member of the profession would behave if he or she were to exhibit each of the twelve ethical traits discussed in the chapter. How do your answers differ from your observations of professionals in these professions?

   a. Sales representative
   b. Doctor
   c. Politician
   d. Corporate CEO
   e. Professor
   f. Lawyer

2. Think of a time in your own life when you felt that there was a perfection to everything. What was it like? What was going on at the time? What brought you to the conclusion that things had happened and were happening exactly as they should? If you cannot think of such a time, what has kept you from doing so?

3. To my knowledge, no one is totally ethical all of the time. Assuming that this is true, for each of the twelve characteristics listed in the chapter, find the nature of the fear that has kept you from expressing that quality. Try to pinpoint the core source of the fear.
4. Which of the characteristics listed do you consider to be the most important and the least important? Why?
5. Try an experiment in humility. On a comfortable night, go to a quiet place outside where you can look up and view the sky. Think for a while of the vastness of that starry universe and the perfection with which it operates. Try to put yourself in other places "out there" and get a feeling for what it must be like in a place where no one ever heard of our planet. How vast is this universe? Think about how we are simultaneously insignificant in the scheme of things and the center of the whole universe. Consider your place in the scheme of things. Then, just sit or lie quietly and listen to the world.

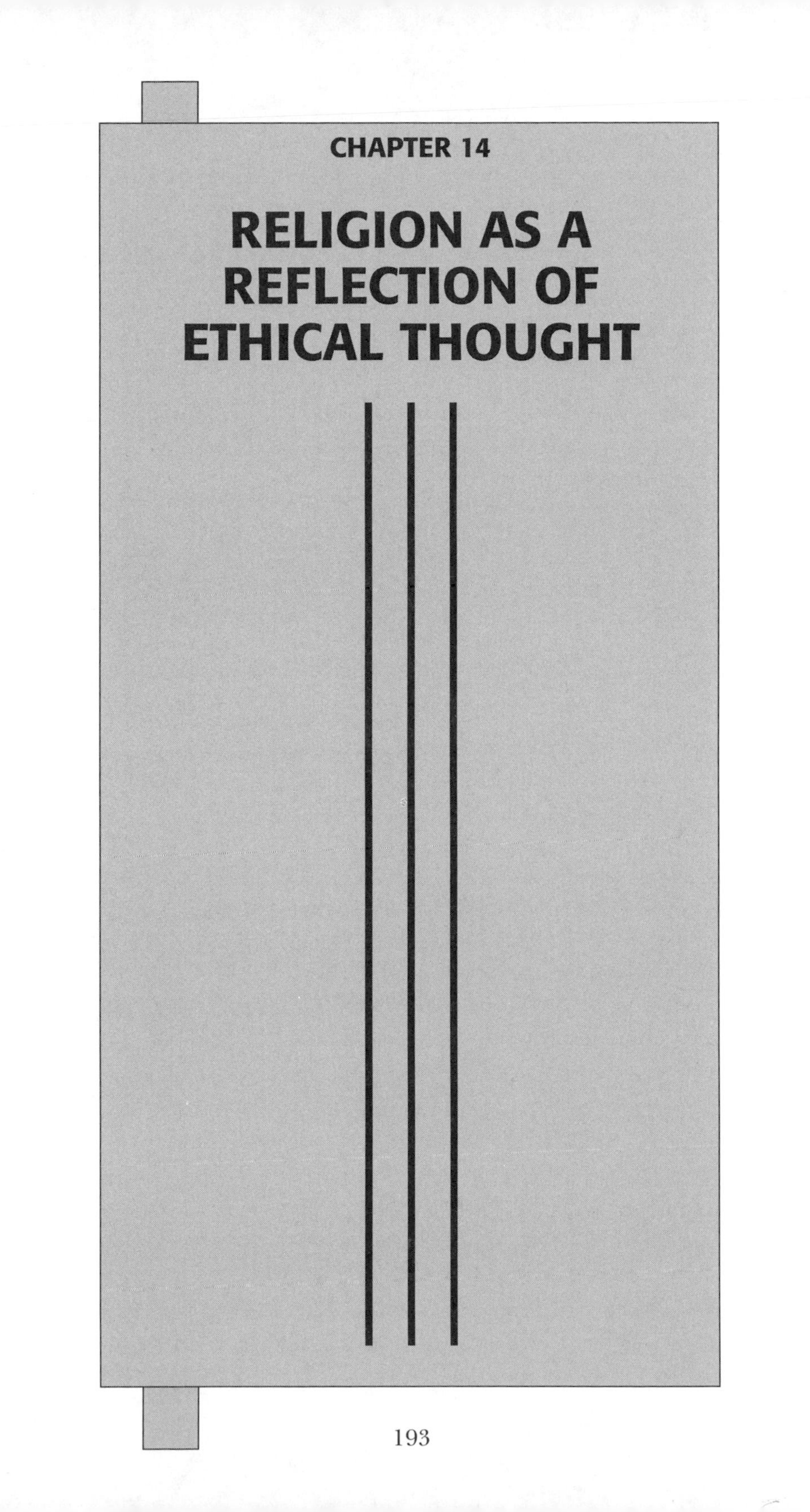

CHAPTER 14

RELIGION AS A REFLECTION OF ETHICAL THOUGHT

INTRODUCTION

At the beginning of this text, I stated that this is not a religious treatise but that religion would appear as part of the presentation. Religion, like myth, tradition, and secular ceremony, is designed to inform the society, to provide information to the members of the society as to what will work for their survival. As such, religion is another method of teaching ethics, at least according to the definition of ethics employed in this book. Religion will be discussed in terms of its ethical content rather than its validity. It is the position of this text that a person's individual religion is acceptable and appropriate, as long as it reflects the truth of what works. There is no intention in this chapter to purport the correctness of one religious tradition over another but rather to show that no matter what religious convictions a person holds, the ethical message is always the same. It is the teaching of this volume that truth is truth, no matter how it is expressed or who says it. We are investigating the essence of religious thought in terms of ethical content rather than the details, myths, belief systems, or admonitions of any one religion.

A representative sample of some of the world's largest religious philosophies includes Judaism, Christianity, Buddhism, and Taoism. The student wishing to verify the propositions offered here for other religions may easily do so once the underlying theme of all religions is established.

JUDAISM AND ETHICS

Although the United States is a culturally diverse nation, consisting of virtually all of the cultural traditions of the modern world, the founding fathers and primary philosophy up to the end of the twentieth century were European, and therefore Judeo-Christian. For this reason, we will begin with an examination of the ethical content of those two religions.

In Judaism, the first five books of the Old Testament, the Pentateuch, represent the historical development of the world from the creation to the founding of the true Jewish culture. These five books include the foundational teachings of the religion and culminate in a series of statements describing the relationship between God and man and among men. These statements are the Ten Commandments, which are the ethical foundation of three of the world's great religions, Judaism, Christianity, and Islam.

The statements are short, concise, and actually quite unambiguous. They represent ten laws handed down by God through Moses to guide the behavior of the Israelites. In essence, they are the statements of ethical behavior required of the chosen people.

If you simply read the Ten Commandments and follow them as statements of law, they become rules to be followed to avoid the wrath of God. This is both religiously and practically appropriate. The emphasis in biblical traditions is to obey the will of God, and therefore one never asks *why* one should be ethical. It is a foregone conclusion as to why you should do it. One should be ethical because that is God's will. But to understand the commandments in a nonspiritual context, we must understand the conditions under which they were presented.

An explanation of the environment in which the laws of Moses and particularly the Ten Commandments were created will be helpful here. Assuming a historical basis for the information of the exodus, though a symbolic interpretation works just as well, we find a patriarchal leader in the person of Moses who is attempting to forge a widely disparate population of related nomadic tribes into a single people and lead them to some as yet unspecified land where they can settle and create their own kingdom. Such an undertaking requires order and cooperation. It is a perfect setting for a religion based on God as master and lord over his chosen people. He is to be obeyed and honored; those who do not follow his rules and regulations (as presented to the people of Israel through Moses) are punished, and those who do are rewarded. Reward and punishment are the essence of the process, and God is the master. As part of this fusing of diverse peoples within the twelve tribes, the commandments are presented as the law of God.

One does not ask why these rules should be followed. One does not question whether any particular one is more important than the others or quibble about meaning and degree of punishment. It is quite simple. The statements say what to do, and the punishment for disobeying is generally damnation. That is a very direct approach to determining ethical behavior.

Yet if we look at each of the commandments from a purely practical point of view (i.e., does it work?), we find that the ten statements are just plain good advice, whether our guide is the law of systemics or the word of God. Consider in each of the ten the principle of reciprocity. Even if they were only suggestions rather than commandments, the rational value of following these tenets becomes obvious, as we find in Exodus 19:14-25:

1. *I am the Lord Thy God, which have brought thee out of the land of Egypt, out of the house of bondage. Thou shalt have no other gods before me.* From a symbolic rather than religious standpoint, this statement can be understood to mean that there is a single guiding principle of life, that that principle is never false, and that thus there is no further need to look for alternative interpretations of how the universe operates. It is this perfect guiding principle that has created the escape from the bondage of ignorance in a foreign reality (Egypt) to the light of truth. The principle in question is reciprocity.

2. *Thou shalt not make unto thee any graven image, or any likeness of anything that is in heaven above, or that is in the earth beneath, or that is in the water under the earth.* From our earlier discussions of the nature of reality, we find no need to put any stock in those elements of what we call the physical world that we observe. The system is too complicated for us to determine the true nature of relationships among physical things and how they are involved in the cause-and-effect nature of our experience of life. This being true, why bother to endow any physical "thing" with imagined powers? The first commandment points out that it is useless, unnecessary, and counter to the truth of the guiding principle of reciprocity. Worshipping idols is a useless process, since this is neither where the power of creation lies nor useful in bringing about the desired results in our lives.

The commandment goes on to say, *Thou shalt not bow down thyself to them, nor serve them; for I the Lord thy God am a jealous God, visiting the inequity of the fathers upon the children unto the third and fourth generation of them that hate me.* Note that there are several things in this passage that serve to make it practically true. Bowing down to idols is, as just stated, useless and unfruitful because the power of creation and truth do not lie in them. Hence, don't bother. It then says that *I the Lord thy God am a jealous God, visiting the inequity of the fathers upon the children.* In terms of a guiding principle, this is not a matter of personal jealousy; it is simply that if a system is to be efficient, it needs to have constant rules of operation that do not vary. That is, the laws of the universe must be the same under all circumstances if it is to be a workable system, and alternative rules are not only unnecessary but also counterproductive.

As far as the punishment is concerned, that is again a matter of reciprocity. If you worship principles or forces that are not real, you receive consequences in kind. Furthermore, the negative consequences of trying to manipulate the world with false assumptions just exacerbates a bad situation and makes things worse over and over again, even to the third and fourth generation. Even the reference to the generations is significant. Consider what happens when you actually do make a mistake in judgment about how things work. You make an assumption and the results of acting on that assumption turn out to be negative. This is an incident. It proves nothing except that you didn't get the result you wanted. This gives no information beyond a specific event. You try again using the same false assumptions, and the results are again unsuccessful. Now you have a coincidence. It still does not "prove" that your assumptions are wrong. If you try it a third time, however, and it still doesn't work, you begin to see a pattern that says these assumptions are false and do not work. Perhaps you should try something else. Occasionally, you might persist in your dogmatic egotism to try one more time using the incorrect assumptions, particularly if you just can't accept the idea of being wrong, and you again fail. By this time, you get the message. Hence, error-thinking leads to failure three of four times, even *unto the third and fourth generation.*

Finally, there is the matter of the phrase *them that hate me*. Error-thinking, operating on the basis of false premises, can be seen as a hateful act if we understand the meaning of the word *hate.* Hate is simply a matter of love held in rather than expressed. As has been mentioned before, if we do not show our love and refuse to express it, through holding it in, we express hate. This need not be an overt act designed to create injury. It is simply a refusal to accept the truth of our existence, and certainly refusing to admit the truth of the guiding principles would do that. The hate, of course, comes back to us as easily as any other act or attitude, and we experience failure as a result of our embracing ideas that simply do not work. Keeping all that in mind, does it seem that this is a command to be followed mindlessly out of obedience or rather out of an understanding that it works in your life?

3. *Thou shalt not take the name of the Lord thy God in vain, for the Lord will not hold him guiltless that taketh his name in vain.* Taking the name of the Lord thy God in vain is merely espousing the power of the guiding principles without believing them. It is literally in vain because if a person purports to know the truth and does not follow it, it does him or her no good. I can speak of reciprocity all I want. I can profess the importance of love as a principle that has value and that I am the creator of my experience of life, but if I do not act in accordance with the principles that enunciate the nature of the universe, it does me no good. I am still going to experience the results of my behavior, not my words.

4. *Remember the sabbath day, to keep it holy.* The sabbath, the one day in seven set aside for religion in Judaism, is reserved for non-worldly pursuits. This reflects working principles because it says that we are not only physical and mental creatures, elements of our nature that predominate for the other six days, but spiritual creatures as well, and this is true whether one wishes to believe it or not. As with other truths, not believing it does not alleviate us from receiving the consequences of that truth. The spiritual nature is that part that best realizes and comes to know the nature of how the system works. This idea of taking one day in seven as a day of rest and meditation not only rejuvenates the physical and mental processes but also serves to remind and strengthen our awareness of the truth. We are reminded of how the universe works so that we do not slip back into decisions and processes that are counterproductive to our ends. It is simply a matter of balance in the threefold nature of our being. Without the balance, we do not function effectively and have limited success in whatever we do. Too much physicality, mental manipulation, or spiritual development at the expense of either of the other two leads to imbalance and disaster. This principle of balance is espoused in fields from religion to psychology.

So far we have viewed the first four commandments, those dealing with our relationship to understanding the nature of how the universe functions. The nature of the relationship between God and humans is the nature of the relationship between universal law and our use of it. In these four statements of action, how best to acknowledge and use this relationship is presented. In the next six commandments, we find an exploration of successful relationships among humans. Understanding how they represent ethical behavior is far easier.

5. *Honor thy father and mother; that thy days may be long upon the land which the Lord thy God giveth thee.* At the time these words were first presented, it was a very practical piece of advice. A child who did not obey his or her parents in a culture that was nomadic, in a hostile environment, and living close to nature would probably not survive for very long. Parents were and are the chief source of information about how to survive. Under the relatively primitive circumstances faced by the tribes of Israel, not listening and doing what one was told could very easily lead to disaster. The message is clear: Listen to your elders; they can teach you how to survive in a hostile world. It is a simple message and to a large extent as true today as it was in the past. Nearly all people rebel against their parents from time to time. It is a natural inclination that comes with growing up and seeking to become wholly functional adults. But those who are truly rebellious will find themselves at risk much more often than those who are so only occasionally. Listening to the wisdom of elders is just plain good advice. Further, since reciprocity is in place, another excellent reason to honor your parents is that when you become parents, your children will honor and believe you as you did your parents. Payback can be hell, particularly if it is the natural kind.

6. *Thou shalt not kill.* Reciprocity is absolute. If you kill, you will probably be killed. Furthermore, if we are all connected and, thus, what happens to one affects all (Shea, 1988), then killing is a form of suicide. The suicide can be seen literally—we set ourselves up for death—or symbolically—we kill a bit of our humanity and connectedness to others in the process, thus killing a part of ourselves. We place ourselves beyond the pale. Little more needs to be said.

7. *Thou shalt not commit adultery.* The admonition against adultery is a very practical one. From the standpoint of the survival of the species, humans operate most successfully by founding family groups within tribal groups and thus dealing with the problems of life collectively. This is a very effective mechanism, both because of the concept of safety in numbers (the tribal element) and because in order for the next generation to come to maturity and continue the species through reproduction, it must be nurtured for an extended period of time. This requires both a mother and a father. The pair bond created in the process of choosing a mate must be maintained for years in order to ensure survival of the child. This being true, it is counterproductive for the male or the female to be involved with other members of the group in ways that destroy this unit. That is the fundamental biological process involved, though higher civilizations, by reducing the need for a maintained pair bond to ensure the survival of offspring, sometimes create tension among opposing tendencies.

Beyond this there is another problem with breaking the rules of monogamy within the society, and that has to do with personal integrity. In this issue, we are back to reciprocity and the natural consequences of our actions. In most societies, there is some form of ceremonial affirmation of the pair bond, resulting in agreements of marriage. Partners take oaths or make promises as to how they agree to behave toward and with each other from the moment of commitment on. If these oaths are broken, there is a breech in integrity. In other words, the parties have broken their word and have become liars. This has serious consequences in a person's life in that (a) if I lie, I will find others lying to me, (b) if I break my word, others will break their word to me, (c) if I act without integrity and honor (respect for others and their rights), then people will find me without honor and not believe me, and (d) if I perpetrate an act of violence by causing injury to another person, I thus injure myself by injuring them. This is obviously not a good idea. Adultery in almost all cases simply does not appear to work.

In fairness, it should be noted that oaths are taken based on available information; if new information reveals that the agreements made do not work for the benefit of all involved, changing our minds and developing new relationships/agreements is an ethical act. It must be done, however, openly and honestly to maintain integrity; all parties must be aware of the new agreements. To be truly ethical, changing agreements will generally be bilateral, with both parties realizing the value of the change and consenting to break the old one. Adultery, being a matter of sexual and emotional attachment to someone other than a spouse, is not an example of changing agreements in that it tends to jeopardize the well-being of the family as a whole.

8. *Thou shalt not steal.* If I steal, I will find thieves in my life or otherwise experience a loss of equal or greater value. Let's be rational here. If theft really worked, then everyone would be doing it and profiting from it. We do not as a group refrain from stealing because it is against the law or because it makes us good people. We do it because it does not work and because embracing such an idea at the societal level would lead to complete chaos and instability. Certainly in Western society and other highly successful market-based economic systems, the sanctity of private property has proven its value over and over again.

9. *Thou shalt not bear false witness against thy neighbor.* Here again is a matter of reciprocity. If I bear false witness against a neighbor by falsely accusing or by gossiping about him or her, I will find that I am falsely accused and will be the subject of gossip. I will additionally receive the brunt of the pain and misery that is caused by such behavior. This is easily understood in practical terms, since false information leads to false conclusions, and we may take all the proper actions in a situation, but if it is based on a false understanding of the situation, it will not work for our benefit. In order to get to where we are going, we must first know where we are; this is true far beyond merely navigating the physical world. In terms of gossip, it must be remembered that gossip is any discussion among two or more parties of third parties *to no one's benefit.* Repeating rumors or sharing information that we do not personally know to be true and that is not for the express purpose of benefiting the people involved is an act of gossip. It creates false realities and becomes very expensive to the practitioner. There is quite enough misinformation in our daily lives to sort through without creating more for our own amusement or aggrandizement. In essence, it simply does not create positive benefit.

10. *Thou shalt not covet thy neighbor's house, thou shalt not covet thy neighbor's wife, nor his manservant, nor his maidservant, nor his ox, nor his ass, nor any thing that is thy neighbor's.* This is probably the most important of the commandment statements because it speaks of the results of acting out of fear rather than out of love. Coveting is a direct refusal to accept love and admit that there is enough to go around. It creates a reality in which the world is one of scarcity, limitation, and shortage; if someone else has something, it means that there is less for you. The person coveting is acting out of fear. Why do I want my neighbor's ox or house or maidservant? Because there isn't enough of those things for everyone to have; therefore, if my neighbor has them, I cannot. This is a philosophy of limitation, the exact opposite of the truth.

Reality is different. Because of the principle of reciprocity and because we create our own realities unconsciously and consciously through our belief systems, there is an infinite supply of whatever we want. If I lack, it is because of my poverty consciousness. If I am afraid I cannot have or keep what I want, it is because I do not understand how it came into existence or how it continues in existence. Nothing is real except as we give it reality. Once this is realized, simply working toward what we want in our lives leads to abundance, and seeing to it that others have what they need leads to us having what we need. It is paradoxical only if we believe in limitation; limitation is a concept of fear. All this has been said before, yet it needs saying again in this context because it is the truth of reality. Coveting is totally unnecessary. Other's wealth is nothing more than an example of what we may have if we so decide. That one shift in consciousness will lead to all the abundance anyone could ever desire.

CHRISTIANITY AND ETHICS

Do not confuse Christianity and ethics with the concept of the Christian ethic. They are two different processes altogether. The Christian ethic is a rather puritanical concept that was developed largely from the teachings of Saint Thomas Aquinas in his apology (defense) of why Christians should follow the rules of the bishops as laid out at the first diet of Nicaea in 325 A.D. Aquinas's document lays out a logical argument for the dominance of the Catholic (universal) church as the authority on Christian doctrine and practices; it centers on the concept of predestination, a view that is antithetical to the presentation of individual responsibility and self-determination that appears in this text. The result of the Aquinas apology is what is known as the Christian (or Protestant) ethic, a puritanical view of the world that effectively says that if you're very good and work very hard in life, you have an even chance of going to heaven rather than hell—though religious pundits may attack that analysis from their individual belief system perspectives.

In discussing Christianity and ethics, the purpose is to show how the teachings of Christianity and the New Testament (the teachings of Jesus of Nazareth) reflect an ethical point of view. Again, we are not looking at this process from a religious point of view but rather from the point of view of how the teachings reflect the principles of what works.

The teachings of the New Testament differ philosophically from those of the Old Testament in that the view of God presented by Jesus is a view of a loving and patient father rather than that of a powerful ruler. God is seen not so much as the Lord God but rather as God the Father. There is a different feel to the concept altogether. With a lord or ruler, the obligations of the people are to obey and do the lord's will. If we change our perspective to that of a father, the image is of a person who is wise, loving, provides what is needed, guides his children, and protects them. The message of the New Testament, or new arrangement, is exactly this paternal message in intent and in content.

The most direct statement of the tenets of Christianity can be found in Matthew V, VI, and VII. These represent the most direct and complete presentation of the philosophy of Jesus and the essence of all his teachings and demonstrations of those teachings. There is a great deal of material here, and it is not the author's intention to go through each line and discuss what it says. However, some comments regarding representative passages will serve well to illustrate the ethical message of the presentation.

Book V of Matthew begins with the Beatitudes:

> *Blessed are the poor in spirit, for theirs is the kingdom of heaven.*
>
> *Blessed are they that mourn, for they shall be comforted.*
>
> *Blessed are the meek, for they shall inherit the earth.*
>
> *Blessed are they which do hunger and thirst after righteousness, for they shall be filled.*

Blessed are the merciful, for they shall obtain mercy.

Blessed are the pure in heart, for they shall see God.

Blessed are the peacemakers, for they shall be called the children of God.

Blessed are they which are persecuted for righteousness' sake, for theirs is the kingdom of heaven.

Blessed are ye, when men shall revile you and persecute you, and shall say all manner of evil against you falsely, for my sake.

Though it may not appear so on first inspection, there is much practical advice here, all of it reflecting a single principle, that of reciprocity. Notice that each statement consists of two parts, one that expresses an action on the part of the individual and another that states the results of that behavior. To understand how this works, we need to consider some differences in language.

Being poor in spirit has nothing to do with physical wealth, nor does it indicate that you should have a spirit of limited extent or poor quality. It is saying that those who spiritually see themselves as a humble part of the greater whole, neither above nor below anyone else, will have a better grasp of the true nature of the universe and dwell easily in a world without fear or want (the kingdom of heaven).

Mourning is an act of completion. It represents the act of remembering and admitting that the absence of someone who has passed away is painful. It is a matter of being willing to feel your feelings rather than bury or sublimate them, and any psychiatrist will tell you that feelings held in tend to fester and manifest themselves in various ways in a person's life, some socially acceptable and others not. The comfort of mourning comes from a willingness to feel the discomfort of separation, to go through the fear connected to the absence of this person in our lives and the fear of our own mortality that a death represents. Once a feeling is truly experienced, it no longer has a hold on us; we are again free to love openly and honestly the person who is no longer with us. The refusal to do so keeps us from holding the departed person in consciousness in a positive, loving way, and this is a form of dishonoring the person in our memory.

We are told that the meek will inherit the earth. What must be understood is that the term *meek* originally meant disciplined and was often applied to teams of horses, such as those used to pull chariots that had learned to work together in unison and therefore maximize their cooperative efforts. I think that's called synergy. Why do they inherit the earth? Because they, through their cooperation, are in a much better position to deal with the unexpected events of life and handle them effectively. Discipline is a very valuable trait in anyone, because it greatly simplifies the process of life and reduces the amount of time and energy necessary to do what is needed.

Those who hunger and thirst after righteousness (are about the business of discovering and mastering the truth about how things work) shall be filled (will succeed). They are receiving what they put into the process. The more they seek the truth, the more the truth is available to them.

The merciful obtain mercy. We are all paid in kind for our behavior.

The pure of heart live without fear and therefore are able to grasp the love that is the universe, a process of seeing God.

The peacemakers are the children of God, because peace is an absence of violence, which is an affirmation of the unity of which we are all a part. This being the case, those who pursue peace are going about the process of seeing themselves as cocreators of the universe and a part of the universal whole, a concept that has already been expressed.

We are not blessed because of the misery of persecution. No one is saying that we should all try to go out and suffer. What is meant is that if an individual is persecuted for telling the truth, that person will live in a world in which the persecution does not matter. An example of this is the individual who is ridiculed for stating unpopular truths. Others may jeer and laugh, but the person is still secure in the knowledge of the truth; the reaction to those persecuting him or her is not one of fear but rather one of compassion for the ignorance of others. Think of how you feel about someone who does something that you know doesn't work, and when you point it out, he or she laughs or becomes angry. Your choice is to become angry or to feel compassion for the pain such individuals will cause themselves. If you truly understand the truth of the situation, compassion is what follows. In each of these cases, persons receive what they give, and the reciprocity is obvious.

Elsewhere in the Sermon on the Mount, Jesus says:

> *Ye have heard that it hath been said, An eye for an eye, and a tooth for a tooth:*
>
> *But I say unto you, That ye resist not evil, but whosoever shall smite thee on thy right cheek, turn to him the other also.*
>
> *And if any man will sue thee at the law, and take away thy coat, let him have thy cloak also.*
>
> *And whoever shall compel thee to go a mile, go with him twain.*

Give to him that asketh thee, and from him that would borrow of thee turn not thou away.

These are strange ideas. It sounds as if the message is to give in to evil and let people steal from you. This is of course not the message. If we remember the ethical principle involved, the ideas make perfect sense. Any attempt to resist evil is an act of defense, which implies that defending yourself is necessary. It is an act of fear. In truth, there is nothing to fear if your actions are ethical and a matter of doing what works, and events that seem negative may actually be positive in the long run. We cannot be harmed by the actions of others. To believe otherwise is an illusion. We can only be harmed by our own actions and the results of those actions. If pain comes our way, we have created it; the external perpetrator is no more than an instrument of our own reward. The paradox here lies in the assumption that something is being done to us rather than by us through others. As for being sued unjustly and giving more than is required, the principle here is one of returning evil with kindness or, rather, reacting to the violence of others nonviolently and with kindness. Rather than a need to defend ourselves, we are presented with an opportunity to offer others an example of what works, to increase our own well-being and possibly be of service to others in changing their consciousness to a more workable way of life. All teaching is by example, and to return the example of violence with an example of violence only serves to perpetuate the fear in the other party.

Further, the passage says give to those who ask. That's simple enough, particularly considering that the reciprocity of the situation returns to you in kind many times over. All of this passage is practical, ethical, and, when understood in the light of universal law, very rational.

A third example of the expression of natural law in the Sermon on the Mount involves the Lord's Prayer. It is not my intention to analyze this passage in detail. To do so would require far too much space. What I do suggest is that the reader think about the content of the prayer in terms of a world of abundance where what we experience is the result of our own perceptions and actions and where reciprocity is absolute. In such a world, for instance, our errors are forgiven in the same manner as we forgive the errors of others in their actions with us. In such a world, the universal principles decide the course of events, not our own egocentric desires, and if we understand those laws, we can create whatever we wish by using them. In such a world, we automatically receive what we need each day (daily bread) in terms of opportunities to give and to learn from our mistakes and our appropriate behavior and in terms of gaining more understanding of how we cocreate the world we live in moment to moment. The entire document is a representation of ethical truth.

Finally, let us briefly examine the prime directive of Christianity, the instruction to *love thy neighbor as thyself.* This is clearly rational and ethical. If we treat our neighbors in a manner that is identical to how we wish to be treated, they will treat us as we wish to be treated. Our actions are returned in kind! Further, if we accept that we are all created in the image of God (all part of the same process) and have a spark of the divine within us (the Christ within), then the New Testament charge to love is nothing more than an insistence to love ourselves and in the process love God as the creator. Again we see that the religious thought and the principles of ethical behavior coincide and that this religious philosophy, at its roots, tells the same truth as others.

BUDDHISM AND ETHICS

Unlike Judaism, Christianity, and Islam, Buddhism (Burtt, 1955) is not a messianic religion in that it does not contain the concept of the savior who has or will come to deliver people from their sins. Assuredly, it does have a prophet, or teacher, in the person of Gautama Buddha, known as the compassionate Buddha, but rather than being a savior, he is viewed as a saint or prophet who teaches a mode of behavior that will lead to spiritual reward and relief from the misery of this world. The primary teachings of Buddhism center on two sets of statements, the first known as the Four Great Truths and the second as the Eightfold Path. In these teachings, the Buddha shows the student the nature of the world and how to escape from its evils. In doing so, the teachings also indicate all that we have said to this point concerning the nature of reality and how to use the universal laws to build a life that works.

The Four Great Truths

According to the Four Great Truths (Burtt, 1955), the nature of life is as follows:

> *Existence is unhappiness.*
> *Unhappiness is caused by selfish craving.*
> *Selfish craving can be destroyed.*
> *To destroy selfish craving, you can follow the Eightfold Path.*

Presented here is an analysis of humanity's misery, beginning with the idea that people are unhappy. That is, they do not have peace of mind and are in various inferior states of health, wealth, self-expression, and so on. The Buddha further states that the reason for this unhappiness is because people have selfish desires (cravings). In terms of what we have been saying, people do not realize the true nature of the universe and therefore have fears centered around insecurity and the belief that there is not enough to go around. They experience lack, mostly because they have desires that are only for themselves, in isolation from those around them, and are searching for ways to personally gain what they desire in the physical world. Because they find value in these relatively valueless commodities and experiences, they are forced to suffer from their assumption that what they need is unobtainable. Cravings are by their nature expressions of lack. This is a false concept and will naturally not work for true happiness; thus people are unhappy.

The third statement assures us that this state of unhappiness is not necessary, that there is an alternative view of the world that will remove the cravings and thus the unhappiness. Note that it is not stated that you should not want the items craved or that they cannot be obtained. What is said is that the unhappiness is unnecessary and that the craving (covetousness, fear of not having enough) can be removed. From the point of view of universal law, there is plenty to go around; lack is not a result of scarcity but rather of the mind-state in which the individual is dwelling.

This brings us to the fourth statement, which is that the shift in mind-state necessary for doing away with the ego-desires is to follow the Eightfold Path.

The Eightfold Path

Each of the elements of the Eightfold Path specifies behaviors that lead to extinguishing cravings. Each, if viewed in light of our discussion of principles in this text, makes perfect sense as a step in developing an ethical, workable life.

1. *Right understanding.* Right understanding is a correct understanding of the nature of the universe and how it functions. From the ethical point of view, this would include being personally responsible, knowing yourself as the creator of your own experience in life, and understanding that all things are possible simply and easily by just remembering to be syner-

gistic, that the universe is reciprocal, and that all things will achieve and maintain balance within the fabric of life.

2. *Right purpose.* Right purpose speaks to the fact that it is as much intent as it is action that creates results. If my motivations are selfish (based on cravings), then the results will be reciprocally unfulfilling. If my purpose is one of benefit to all involved, the results will be positive for me. The question is, To what am I aspiring? What is my reason for the action? This is as important as the actions themselves. Alms, for instance, given for the benefit of others create wealth in my life. Alms offered to create a good opinion on the part of others with little regard for the needs of those in need will only result in my own impoverishment.

3. *Right speech.* Right speech has not been addressed much in the text, but it is nevertheless quite important. What we say impacts what we think. If I say something often enough, even in jest, it may result in my mind believing it to be true, and therefore I experience it in my life. Speaking ill of others in anger or in jest is the first step to thinking ill of others, and it becomes my truth. Speech is a valuable tool in creating our world as we want it (without lack). Thus, we should guard our speech so that we feed our mind only positive ideas.

4. *Right conduct.* It is a fairly straightforward concept to understand that our conduct has consequences. This has been much discussed in earlier chapters.

5. *Right vocation.* The principle is not that some occupations or activities are valuable and others are not. It is rather an understanding that some activities and occupations are reflections of our own unique talents and an expression of what we have to offer the world if we are operating at the highest expression of ourselves. That is, some occupations are statements of self-expression, while others are chosen for reasons other than being ourselves, such as to seek money (a craving) or fame (another craving) or because we are told we should (and thus we try to avoid ridicule, being wrong, or making an error). When we choose a correct vocation, we find that we express ourselves freely and easily and that the results are better than with anything else we may have chosen to do. Working at the wrong job only leads to a feeling of emptiness and loss, which, of course, is another expression of selfish want.

6. *Right effort.* The question here is whether our efforts are constructive and serve to satisfy our goals or merely allow us to look as if we are working toward those goals. Meditation is only valuable if it is entered into earnestly and honestly, not half-heartedly. In terms of ethical principles, effort that is directed toward doing what works bears fruit, while effort directed toward selfish fulfillment of only personal cravings leads to the loss of what one already has.
7. *Right alertness.* In order to behave ethically, it is necessary to be aware of your behavior, to monitor it, and to make choices that are to the benefit of all involved. At first this does not come easily. We must be ever vigilant in order to safeguard against digression into fear motivation and attempts to satisfy only personal desires at the expense of others. Awareness, alertness, and vigilance are necessary until right behavior and thought become so engrained and natural that we can make the transition from unconscious ineptness to conscious ineptness to conscious adeptness and finally to unconscious adeptness. It involves recognizing when and how we are in error, learning not to become so, and first consciously and then unconsciously making ethical choices. Until we reach the stage of unconscious adeptness, when being ethical is so engrained that it is done without thinking, we need to remain alert and aware of our actions.
8. *Right concentration.* This final step is the method by which we concentrate our awareness and guard against our own fears. It is perhaps the most difficult of the eight because there are so many choices as to how we may achieve an understanding of the process that until we find one that works, we tend to founder. Ultimately, we find that all that is needed is to concentrate on a small group of principles and to keep them as the determining factors in all our decisions. Those principles have been enunciated repeatedly in this chapter. As mantras, repetitive statements to keep us focused, they are simply "reciprocity is absolute" and "love thy neighbor." If we learn no more than these two, the rest of truth follows naturally.

Of course, the Buddhist religion contains much more in the way of truth. It parallels in its principles very closely the main teachings of the other great religions. Buddha taught and preached widely, and not only his own words but the expanded understanding of other

Buddhist monks make this an incredibly rich religious tradition that extends back to the sixth century B.C. Yet I feel it is valuable to mention one final element of the teachings, that being the last teaching of Buddha, the Lotus Sutra.

In the Lotus Sutra, Buddha points out that all of the teachings that have gone before, all of the methodology and explanation and exercise, are unnecessary to get off the karmic wheel of birth, life, death, and rebirth, an endless progression of unhappiness. All that is truly needed is to realize that the physical universe is fantasy and that we make it all up. Once that is truly realized, there is no more fear, no more lack, no more scarcity. There is only a vast universe to play in and infinite opportunities for exploration of who and what we are.

TAOISM AND ETHICS

Taoism (Lao-tzu, trans. 1988) is an Eastern or Oriental religion/philosophy that is nonmessianic and represents more a format for daily behavior than that which a Westerner would call a religion. It is expounded in most detail in the *Tao Te Ching* (Teaching of Tao) and is attributed to Lao-tzu, an apparent contemporary of Confucius in the sixth century B.C. More likely, it is the collected wisdom of a larger number of sages. Whatever the case, *Tao Te Ching* is beautifully and lucidly written and contains volumes of knowledge in its short, precisely formulated verses.

The study of the *Tao* is the study of The Way, the road that leads to a perfect expression of self without flaw. To achieve such a state is to act appropriately and in harmony with all of life without having to think about it. (This is very close to the transcendental ethics purported as the highest state of ethical development.) To the Taoist, it is extremely difficult to achieve, to the point of being practically impossible. Yet it is the goal of the process of a Taoist life. In essence, the person seeks to achieve a state of perfection of action, thought, and being, while knowing full well that it cannot be achieved. Thus, the first paradox arises. In truth, achieving the state is not the goal; rather the goal is the journey. No matter how much individuals concentrate on achieving perfection, they will not attain it, but as they move down the road toward perfection by following the Way (method, path, behavior), they find themselves at an ever higher level of understanding. Their lives become gentler, simpler, ever more perfect, and others recognize them as operating at a higher level than those around them. At the same time, they are not attempting to do any of this. They are simply working on their own perfection.

The general approach of the Taoist is one of *wei-wu-wei* or, literally, doing-not-doing. As paradoxical as the concept of doing by not doing, the presentations of the work often appear to be paradoxical in nature as well. Yet it is in the unraveling of the paradoxes that one is able to gain knowledge. This is similar to a Buddhist koan. However, the koan purposely has no logical meaning, and the verses of the *Tao Te Ching* actually do, once they are understood.

Earlier it was said that all paradoxes are only apparent, and by virtue of that, if we merely change the way we view what is going on, we gain understanding and the paradox disappears. Such is the case with the writings of Lao-tzu. Some samples will serve well to illustrate not only this point but also the ethical nature of what is taught. Considering the clear nature of the passages, it is quite unnecessary to comment on them or explain how they relate to the ethical principles we have discussed. The *Tao Te Ching* is the clearest expression of how an ethical person behaves that I have found. These example verses have been adjusted for modern language and understanding, but otherwise, they are as they were originally presented. I leave it to the reader to discover their meaning.

View the world as yourself.
See the perfection in the way things are.
Love the world as you love yourself;
Then you can care for all things.

The Master of the Tao watches the natural world,
But only trusts his inner senses.
He allows things to come and go
His essence is open to everything.

The Master does not speak the Truth, he Performs the Truth.
When he has completed his tasks,
The people say, "We are amazed:
We have done this thing on our own!"

When people forget the Way,
The concepts of Piety and Goodness arise.
When we lose contact with our natural wisdom,
We turn to cleverness and logic.

When there is strife in the family,
We become sanctimonious.
When the country falls into chaos,
We turn to patriotism as our guide.
The Way to wholeness is through fragmentation.
The Way to structure is through chaos.
The Way to fullness is through emptiness.
The Way to rebirth is through death.
The Way to have all is to give all away.
The Master sees the Truth of all things,
And is content with their being.
She lets them be as they are,
And remains in the center of the circle.
She who remains centered in the Way
Goes wherever she wishes without fear or harm.
She sees only universal harmony in life,
Even where there is pain,
Because she is at one with herself and with the Tao.
The Tao cannot be touched, felt or otherwise sensed,
Yet it feeds and completes all that is.

COUNTERPOINT AND APPLICATION

There must be a single religion that best explains the reality of the world in which we live. The competing religions are misconceptions, pagan or heretical misunderstandings, that should be supplanted by the One True Religion. Because of this, it really doesn't matter what other religions say. Isn't that so?

If you wish it to be so, then it is so. Be aware that it is your reality, and you are welcome to create it however you wish. Whatever your religion, from the standpoint of ethics, it is the correct one, because its basic tenets are identical to those taught by all other religions. It is only the individual interpretation of the details that differ, and that is a matter of personal choice. If there is any caveat it is that the choosing be done in the light of your own personal experience and intuitive feelings rather than according to what someone else has told you. If you wish to convert others, have at it. The proof of that pudding will be in the eating, and if what you are doing is ethical, it will work. If not, you will find you are spinning your wheels. Find your own inner sense of what is real and who/what God is to you. With that as your guide, any choice of religion or spirituality will be perfect for you.

EXERCISES

1. According to the text, the core ethical message of all the major religions is the same. If that is so, what is the reason for so many different religions? (Hint: Consider this question from the point of view of an anthropologist studying cultures rather than a theologian searching for the truth of the nature of God.)
2. The book offers statements from four religions, two of them messianic, one from the Indian subcontinent, and one that is Asian-philosophical. There are obviously other equally important and powerful religions. Examine several others to see if you can find statements that are similar to the ones offered in the chapter.
3. Since an examination of the natural world and a study of the major religions reach the same conclusions regarding what works, what is the lesson to be learned? There is no correct answer. If you honestly address this question, you will gain insight into your own reality, your own motivations, and the strength of your own convictions. This is a more introspective than intellectual exercise, as befits the subject of the chapter.

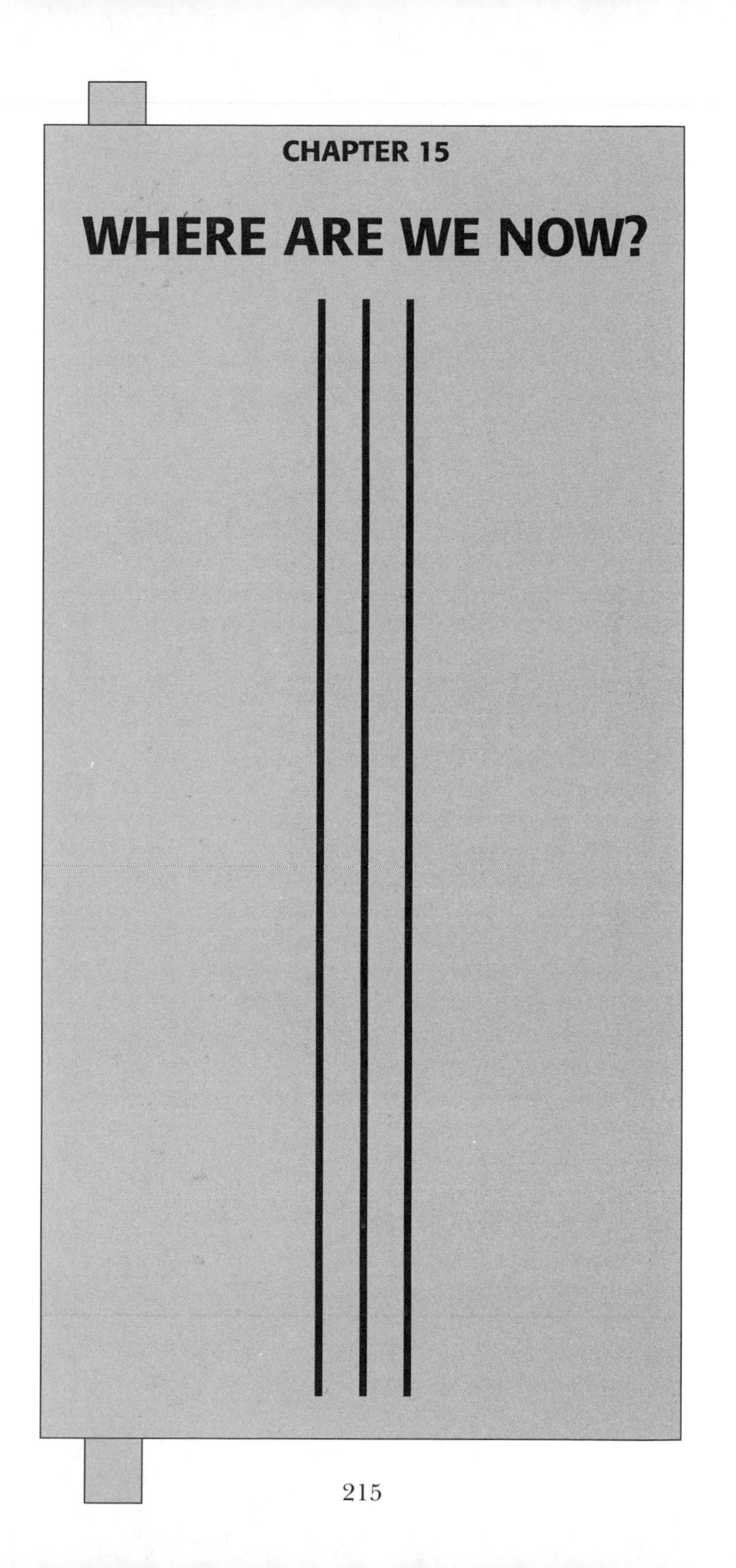

CHAPTER 15

WHERE ARE WE NOW?

Every textbook should have a summary, or a statement of what has been presented, including all of the main points and how they serve to prove the thesis of the work and support the arguments therein. This book is no different, and as with many books that deal with the truth, it has come full circle, bringing us once again to the chore of defining what is ethical and how to behave ethically.

As you may have surmised, defining what is ethical is an ongoing process. It is ongoing not only because individual circumstances dictate what ethical behavior looks like at any given point in time but also because the state of our individual and collective conscious will determine our capacity to see the truth and interpret that truth effectively. Be aware that no one is purposely unethical; we all strive to do what works for us. It is simply that at any given point in time, our ability to perceive what works is limited. As we develop expertise, have new experiences, and learn to rely more and more on our innate value system, we become more and more capable of making decisions that are highly ethical. We also find that the process becomes easier and easier as we progress.

We have tools in this process that help us to smoothly make the transition from unconscious ineptness to conscious ineptness, then to conscious adeptness and finally to unconscious adeptness. We have the information provided by behaviorists, such as Maslow, Piaget, Kohlberg, and Shea, who have observed the nature of successful people who exhibit a highly developed state of peace of mind. We have the work of religions and religious leaders, such as Moses, Jesus, Buddha, and Lao-tzu, not to mention the knowledge and experience of those leaders of religions not used as examples in the text.

We also have the information contained in this book concerning the relationship between love and fear and how each affects not only our behavior but also our level of well-being. We have discussed the individual nature of reality, the laws by which this systemically oriented universe functions, and the fact that these laws—of synergy, reciprocity, and balance—are immutable. And we have discussed that by living in the Self state, we can all have health, wealth, love, and perfect self-expression, which lead to peace of mind.

We have seen how the experience of our lives is our own choice and that in reality it is only our tendency to defend ourselves when we are afraid that motivates us to ever carry out an unethical act, and that to do so results in our own misery. Ah, sweet revenge—it truly is always against ourselves.

All of this material has been presented, discussed, and supported by examples and exercises. Yet where does that leave us? How does this prepare us for a more ethical life if it is all just so much rhetoric, so much theory?

In fact, as rhetoric and theory, it does not prepare us for a more ethical life. But as a guide to action, it does. Just as love is an active verb, so ethics is an active concept. It is not something we learn, and it is not something we analyze. It is something we do, and until it is actually put into action, it is a useless concept, nothing more. The secret to this book centers on the fact that it is a *practical* work, designed to be used, not just studied. How often have you been willing to stand on your principles rather than just go along with the crowd or protect your own position in a situation? How often have you followed your heart, no matter what your head told you the "logical" thing to do might be? When is the last time that you actually stopped and thought about your normal, habitual mode of behavior as compared with the behavior of someone with any of the twelve traits of an ethical person? It is my hope that this has happened more and more as you have progressed through this book; if not, then the book is useless to you. It is nothing more than so much ink and paper. The important issue here is not the book itself or how it presents its message. The important thing is whether or not you have understood the message or *principles* discussed in the book, whether or not you have begun to incorporate that message into your life.

Make no mistake about the purpose of this work. It is designed to institute a change, in most cases a major one, in the manner in which the individual readers lead their lives. The purpose of this book is to be a subversive force in your life that will eat at you until you begin to see that there is a more effective way to live life, that all of those unhappy events, and loss, and pain, and loneliness are totally unnecessary. The purpose of this book is to wake you up to the fact that you do not have to live in a state of disease, poverty, isolation, and dissatisfaction and that a very large part of the nature of your experience lies in your choices.

Additionally, the purpose of this book is to specifically and directly address the relationship between our choices ethically and our actions technologically. Technology is still our chief source of control in the physical world. We are better at it than any other creature on the planet; it allows us to dominate within the limits of our desire to dominate as we control, restructure, and affect our environment and all living things in that environment. It is a wonderful and powerful tool, and with the gift of technology, we also, consciously or unwittingly, accept the obligation of that power because our actions affect everything around us.

We have a tremendous capacity to contribute to the well-being of every living thing on this planet. We do in fact alter the entire system with our choices, and we are consciously aware that we do not know all of the consequences our technological choices will produce. Yet most people involved in the process mindlessly push on, ignoring those consequences or the implications of the overall order within the system. Most of us choose to forget that for every action there is an equal and opposite reaction and that, whether we like it or not, the decisions we make will return to us in kind. If those choices are considerate of results and take into account how they will affect others, we will be treated with consideration in return. If our choices are to ignore others, then we will be ignored in return, by other people, by the environment, and by the machinations of the entire planet.

Make no mistake about it. We are not in charge in the way we believe we are. We alter nature to our benefit because we are allowed to, not because we are in charge. When we make mistakes in judgment regarding how we use technology, we experience the results of those choices. Those results may include drought, massive lead poisoning, the demise of multiple species, the sterility of land, the displacement of whole societies, plagues, pollution, the loss of ocean fecundity, global warming, and a world in rebellion. They may conversely include a cleaner environment, improvement in the health of the population, and peace and prosperity. One thing is certain. The system will adjust itself. This is a basic law of nature. We are part of that nature and part of that adjustment as the system constantly reestablishes balances. And that will continue, but only as long as the existence of our species is useful to the overall survival of the system itself. Systems that do not change and adapt cease to exist through a lack of synergy. If we become nonsynergistic and do not reestablish a synergistic relationship with the whole, we will be sloughed off as easily as we sweep an insect from our arm.

Doing what works is doing what works for the entire system, and that requires discipline and forethought. It is here that we must learn to excel, so that our decisions are positive ones, even to the seventh generation. It is here that we must begin to see the manner in which our actions affect others and to choose in the world of technology in a positive manner, just as we do in every other aspect of our lives.

It is our responsibility. We are the ones with the ability to respond. And if we exercise that capacity legitimately, for the benefit of all, in a loving and fearless manner, the results cannot help but be positive for ourselves and for the rest of humanity. Think about the choices you make. Do you consider more than your own needs? How many people consider more than their own selfish needs? And what are the results of those who do and those who do not? Know yourself by your fruits, by the results of your actions, and let them be a guide. Wisdom is gained only through experience. It is the culmination of knowledge and love in combination. Gain wisdom and become healthy, wealthy, and capable of perfect self-expression and of giving and receiving love. In other words, find your true peace of mind.

Finally, enjoy the journey. We are here for this journey, and every part of it is necessary to our natural unfolding. No effort is wasted and no event is insignificant, if we will but choose to learn from it. Remember that in truth, the universe is a benevolent place; if we will but learn the key, we will find that it will provide more success, more happiness, more contentment, and more adventure than we can imagine. Relax. Enjoy the ride. And learn. If you are truly fortunate, you will find yourself in this school of learning for a very long time.

Bibliography

Alcorn, Paul A. *Social Issues in Technology: A Format for Investigation*. Upper Saddle River, N.J.: Prentice-Hall, 1997.

Allen, James. *As a Man Thinketh.* New York: Grosset & Dunlap, 1983.

Allen, John L., ed. *Environment: 94/95.* 13th ed. Guilford, Conn.: The Dushkin Publishing Group, 1994.

Anonymous. *The Way Out*. Tonawanda, N.Y.: Sun, 1971.

Attenboro, Richard. *Words of Ghandi.* New York: New Market Press.

Bach, Richard. *Illusions: The Adventures of a Reluctant Messiah*. New York: Dell, 1977.

Bertalanffy, Ludwig Von. *General Systems Theory*. New York: George Brazillen, 1968.

Bothamley, Jennifer. *Dictionary of Theories*. London: Gale Research International, 1993.

Burtt, E. A. *The Teachings of the Compassionate Buddha*. New York: Penguin Books, 1955.

Capra, Fritjof. *The Turning Point: Science, Society, and the Rising Culture*. New York: Bantam Books, 1982.

Casti, John L. *Complexification: Explaining a Paradoxical World Through the Science of Surprise.* New York: HarperCollins, 1994.

Crain, W. C. *Theories of Development.* Upper Saddle River, N.J.: Prentice-Hall, 1998.

David-Neal, Alexandra, and Lama Yongden. *The Secret Oral Teachings in Tibetan Buddhist Sects.* Translated by Capt. H. N. M. Hardy. San Francisco: City Lights Books, 1967.

Ellin, Joseph. *Morality and the Meaning of Life: An Introduction to Ethical Theory.* New York: Harcourt Brace, 1995.

Ellis, Peter Berresford. *The Druids*. Grand Rapids, Mich.: William B. Eerdmans, 1994.

Facione, Peter A., Donald Scherer, and Thomas Attis. *Ethics and Society*. 2nd ed. Upper Saddle River, N.J.: Prentice-Hall, 1991.

Foundation for Inner Peace. *A Course in Miracles*. Tiburon, Calif.: Foundation for Inner Peace, 1985.

Fox, Emmet. *The Sermon on the Mount*. New York: Harper & Row, 1934.

Fox, Emmet. *The Ten Commandments*. San Francisco: Harper & Row, 1979.

Freud, Sigmund. *The Future of an Illusion*. Translated by W. D. Robson-Scott. Reprint, Garden City, N.J.: Doubleday, 1964.

Gibran, Kahlil. *Sand and Foam*. New York: Alfred A. Knopf, 1993.

Goldberg, Philip. *The Intuitive Edge: Understanding Intuition and Applying It in Everyday Life*. Los Angeles: Jeremy P. Tarcher, 1983.

Graham, Loren R. *Between Science and Values*. New York: Columbia University Press, 1981.

Harris, Marvin. *Cultural Materialism: The Struggle for a Science of Culture*. New York: Vintage Press, 1979.

Hassan, Ihab. *The Right Promethean Fire: Imagination, Science, and Cultural Change*. Chicago: University of Illinois Press, 1980.

Herm, Gerhard. *The Celts: The People Who Came Out of the Darkness*. New York: St Martin's Press, 1975.

Jones, Judy, and William Wilson. *An Incomplete Education*. New York: Ballantine Books, 1995.

Jordan, James N. *Western Philosophy from Antiquity to the Middle Ages*. New York: Macmillan, 1987.

Kegan, Robert. *The Evolving Self*. Boston: Harvard University Press, 1983.

Keyes, Ken, Jr. *The Hundredth Monkey*. Coos Bay, Oreg.: Vision Books, 1982.

Kohlberg, Lawrence. *The Meaning and Measurement of Moral Development*. N.Y.: Clark University Heinz Werien Institute, 1981.

Kottak, Connrad Phillip. *Cultural Anthropology*. New York: McGraw-Hill, 1994.

Krishnamurti, J. *The Urgency of Change*. Edited by Mary Lutyens. New York: Harper & Row, 1970.

Kuhn, Thomas. *The Structure of Scientific Revolutions*. 2nd ed. Chicago: University of Chicago Press, 1970.

LaFarge, Oliver. *History of the American Indian*. New York: Crown Publishers, 1961.

Lao-tzu. *Tao Te Ching*. Translated by Stephen Mitchell. New York: Harper & Row, 1988.

Martin, Mike W., and Roland Schinzinger. *Ethics in Engineering*. 2nd ed. New York: McGraw-Hill, 1989.

Maslow, Abraham H. *Toward a Psychology of Being*. 2nd ed. New York: Van Nostrand Reinhold, 1968.

Mumford, Lewis. *Technichs and Civilization*. New York: Harcourt Brace, 1963.

Nhat Hanh, Thich. *Living Buddha, Living Christ*. New York: Riverhead Books, 1995.

Pelikan, Jaroslav, ed. *The World Treasure of Modern Religious Thought*. Boston: Little, Brown, 1990.

Piaget, Jean. *Moral Judgement in the Child*. Translated by Margorie Gabain and William Damon. New York: Free Press, 1997.

Richter, Maurice N., Jr. *Technology and Social Complexity*. Albany, N.Y.: State University of New York Press, 1982.

Sahal, Devendra. *Patterns of Technological Innovation*. Reading, Mass.: Addison-Wesley, 1981.

Schoeps, Hans-Juachim. *The Religions of Mankind: Their Origin and Development*. Translated by Richard Winston and Clara Winston. New York: Doubleday, 1966.

Shea, Gordon F. *Practical Ethics*. New York: AMA Membership Publications Division, 1988.

Smith, Sheldon, and Philip D. Young. *Cultural Anthropology: Understanding a World in Transition*. Boston: Allyn and Bacon, 1998.

Solomon, Robert C. *A Handbook for Ethics*. Fort Worth, Tex.: Harcourt Brace, 1996.

Steward, Julian H. *Theory of Culture Change: The Methodology of Multilinear Evolution*. Chicago: University of Illinois Press, 1972.

Wilson, R. Anton. *Illuminati Papers*. Berkley: Anchor Press.

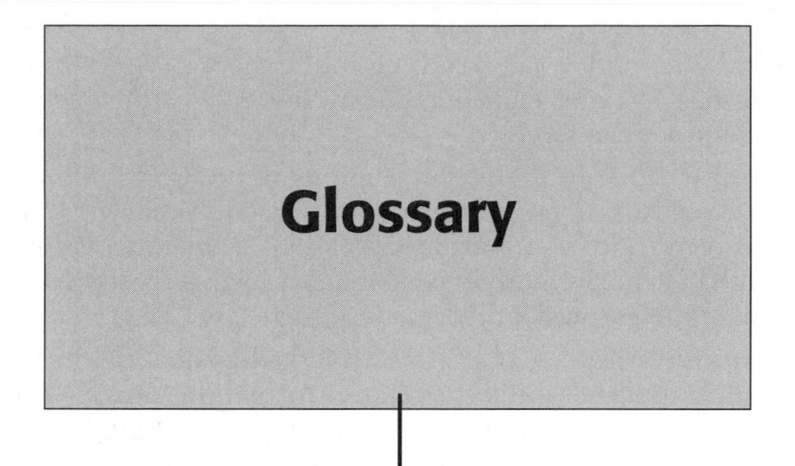

Acculturation A process by which two cultures in contact with each other adopt certain features of the other culture.

Assimilation A process by which members of a culture or entire cultures that enter into another culture take on the characteristics of the larger culture while retaining certain aspects of their own.

Attitude As used in the text, an incorrect perception of reality based on misinterpretation of experiences in the past and upbringing.

Balance The third of the universal laws of systemics. According to the law of balance, all systems are in the process of achieving and maintaining balance (balance is defined by the goal of the system).

Behavioral laws of systemics The three universal laws of systemic behavior governing the manner in which all systems operate. Included are the laws of synergy, reciprocity, and balance.

Boredom A mind state in which an individual is locked in the repetitive behavior of a life without development or growth.

Buddha Lord Gautama, a prince of the Indian subcontinent who founded the Buddhist faith through revelation and meditation.

Buddhism One of the major religions of the world, founded by the Buddha and based on the premises of the Four Great Truths and the Eightfold Path.

Center As used in the text, the representation of balanced existence in which individuals operate from their intuitive sense of appropriate behavior and naturally do what works.

Chaos theory An extension of systems theory, in which the world structure is viewed as both chaotic, without form, and systemic, with order and form. According to the theory, every chaotic system has an order embedded in it, and every ordered system is chaotic at deep levels.

Christianity One of the major religions of the world, founded by Jesus of Nazareth, the Christ, and based on concepts of love and reciprocity as the road to salvation. Its main tenets are presented in the Sermon on the Mount, which is recounted in several of the gospels of the New Testament.

Communicativeness A trait of an ethical individual in which there is an enhanced ability to relay information clearly and succinctly, in a way that is easily understood and contains no subterfuge or avoidance of truth due to fear, and the ability to receive information clearly and unambiguously and understand its total content as well as context.

Complex systems The concept that all systemic structures are complex, with numerous relationships and interactions forming the cooperative network of behavior that allows the system to successfully adjust to change and survive. The more complex the system, the more difficult it is to determine the cause-and-effect relationships taking place within it.

Consideration A trait of an ethical individual in which the actions of that individual reflect an understanding of the needs and conditions of others in such a way that there is every attempt to affect others through those actions in a positive rather than negative manner.

Conventional ethics One of the categories of ethical behavior presented in the theories of John Dewey in which the individual decides what is and is not ethical based on the mores and customs of the society or group to which he or she belongs.

Cooperation Literally, working together. Cooperation represents the synergistic operation of a system in a way that supports efficiency and allows a system to function successfully.

Courage A trait of the ethical person in which that person chooses to act in the face of fear. The interesting characteristic of courage is that it is self-extinguishing; once individuals act in the face of fear, they discover how little there is to fear, and it no longer takes courage to act.

Creativity 1. A human characteristic by which individuals create their world to fit their view of the world. 2. The act of developing new ideas and technology based on our understanding of how the universe operates.

Decisiveness A trait of ethical people by which, through clarity of thought, decisions are made easily and effortlessly, with little possibility of error.

Deterministic randomness The concept of chaos theory that explains why deterministic, cause-and-effect processes in complex systems cannot be predicted. Basically, with sufficient time and resources, an event can be traced forward from its original cause to any given outcome, but a given outcome cannot be traced back to a specific cause because there are so many possible paths that could have been taken to reach that cause.

Dewey, John A philosopher and ethicist who developed a view of ethical behavior that classifies individuals according to the basis upon which they make decisions. The classifications include preconventional ethics, conventional ethics, and postconventional ethics, each with its own determining features.

Dis-ease A state of physical, emotional, mental and/or spiritual ill health resulting from operating in the mind state. It is normally a physical manifestation of mental imbalance.

Druid A priestly class among the Celtic culture of Europe who represented the teachers, religious leaders, and lawgivers of the culture. They were the repository of privileged and arcane knowledge within the culture.

Dynamic system A nonstatic system in which there is a constant adjustment to internal and external conditions, necessitating change on an ongoing basis.

Efficiency The process of getting the most out of a system with a minimum input of time, effort, and resources. This represents the functioning of a system at its most perfect.

Eightfold Path A series of statements in Buddhism of eight modes of behavior that collectively lead to freedom from attachment to worldly pleasures and result in true happiness.

Ethical traits A set of twelve behaviors exhibited by ethical individuals who consciously understand how to do what works. It includes integrity, truthfulness, courage, focus, decisiveness, centeredness, consideration, creativity, communicativeness, acceptance without judgment, awareness of perfection, and humility.

Ethics 1. The study of the rules and customs concerning the acceptable behavior of members of a culture and how to determine right behavior. 2. As used in this text, the study of doing what works.

External self-esteem Self-esteem stemming from the admiration and acknowledgement of others. It is a part of the Maslowian concept of belonging needs, as it enhances the feeling of the individual that he or she has membership and a place in the group.

Fear The opposite of love. Fear is the refusal to admit the perfection of the universe, which results in a defensive posture designed to protect one's self from the dangers of the world.

Focus A trait of the ethical person in which one has the capacity to maintain concentration on what one is doing without distraction or deflection due to attitude or fear.

Four Great Truths A primary concept of Buddhism that represents the revelation of Buddha in which he discovers how the evil and unhappiness of the world comes about. The Four Great Truths include (1) existence is unhappiness, (2) unhappiness is caused by selfish craving, (3) selfish craving can be destroyed, and (4) it can be destroyed by following the Eightfold Path.

Fragmentation A mental condition experienced by those who do not understand the ethical basis of the world's structure by which competing ideas and attitudes stifle decisive action because there is no coherence of thought or belief within the individual.

Gandhi, Mahatma A twentieth-century Indian religious leader and activist who was instrumental in bringing about the independence of India from British rule. He was a major proponent of nonviolent protest and one of the most highly ethical individuals of the last century.

Group membership Part of the Maslowian concept of belonging, in which there is a felt need in all humans to be a member of the group. This is an innate survival trait of the species that tends to operate best in extended families and tribal groups.

Group norm orientation That stage in the Piaget-Kohlberg theory of ethical development in which the individual bases ethical decisions on the behavior and acceptability norms of the group or culture. It is similar to Dewey's concept of conventional morality.

Health A state of being experienced by those not living in the mind state. In this state of physical well-being, a person is balanced and has an understanding of the functioning of the universe. It is in contrast to disease (or dis-ease).

Higher level needs A level of needs that include self-esteem and self-actualization, the two highest need levels in Abraham Maslow's hierarchy of needs theory. They are internal and independent of input or connection with others.

Humility A characteristic of ethical people in which they live in a state of awe at the perfection of the universe. Through clear seeing, it has become obvious to these individuals that the process of life works without flaw, and they are humble in the face of such perfection.

Integrity A quality of ethical people in which individuals say what they mean and mean what they say at all times.

Internal self-esteem One of the higher level needs in Maslow's hierarchy. Internal self-esteem represents the feeling of self-worth and personal value received from realizing the perfection of one's own being. It is a major factor in the humility experienced by truly ethical individuals.

Judaism A Middle Eastern religion having common ancestry with Christianity and Islam in the Abrahamic tradition. One of the world's great religions, its main tenets are found in the first five chapters of the Old Testament and are exemplified by the Ten Commandments.

Karma 1. From the Sanskrit, literally, action. 2. The concept that each person is repaid in kind for all acts; an expression of the concept of reciprocity.

King, Martin Luther, Jr. A twentieth-century African-American leader and supporter of nonviolence as a method of achieving social goals. Much of his philosophy was based on the works of Gandhi and Henry David Thoreau.

Kohlberg, Lawrence An ethicist and behaviorist who developed a theory of ethical development based on the work of Jean Piaget.

Law and order orientation A stage in the Piaget-Kohlberg theory of ethical development. In this stage, ethical decisions are based on the law as a formalized statement of group norms. It is simi lar to Dewey's conventional ethics.

Law of Moses The embodiment of Jewish traditional religious law contained in the first five books of the Bible.

Limitation That which keeps individuals from creating their world the way they would like it to be. It is a false belief that puts limits on possibility and denies the creative capacity of the individual.

Love One of the two motivations for behavior (the other being fear). Love is a natural desire for the well-being of everyone and a feeling of connection, respect, and unity with all others. It is expressed in the Greek word *agape* and should not be confused with sexual love or love of siblings, family, objects, or ideas.

Maslow, Abraham A twentieth-century behavioral psychologist who developed the hierarchy of needs as an explanation for human behavior. It approximates the findings of the behaviorist ethicists.

Mind state A state centered in attitudes of scarcity and fear in which it is necessary to maintain control and solve problems in order to survive.

Morality of cooperation The state of child development as proposed by Jean Piaget in which the child begins to realize that rules are created by individuals and groups for particular reasons and that, likewise, individuals and groups can modify, change, or eliminate those rules by mutual agreement.

Natural selection An evolutionary process in which nature selects survival traits within a species, eliminating behavior antithetical to survival. It is a natural tendency of a species to gather and maintain only those traits and behaviors that contribute to survival as a species.

Paradigm The understanding of the nature of reality held by an individual or group of individuals by which they make decisions and learn to operate in their environment. It tends to set the outside boundaries on what can and cannot take place within their world. All new information is generally filtered through the premises of the individual's or group's paradigm.

Paradox An apparent condition in which two diametrically opposed ideas or sets of conditions exist simultaneously. Once it is understood that it is the lack of knowledge about what is taking place that causes the apparent condition, paradoxes can be resolved by viewing activities with an expanded paradigm.

Peace of mind A general feeling of well-being that stems from a combination of knowledge and acceptance of the perfection of the universe.

Pentateuch The first five books of the Bible containing the Law of Moses and particularly the Ten Commandments.

Perfect self-expression A condition of one living in a state of awareness by which there is an ongoing expression of the truth of one's Self without fear and without hiding any information. It represents the highest order of personal truth.

Personal reward orientation An early stage in the Piaget-Kohlberg theory of ethical development. In this stage, ethical decisions are based on the presence or absence of direct personal reward as a result of an action.

Physiological needs The lowest level of need fulfillment in Maslow's hierarchy of needs theory. It involves the satisfaction of survival requirements such as food, water, and warmth. It is considered primary and most urgently sought by individuals.

Piaget, Jean A twentieth-century developmental psychologist who developed a theory of mental development in children, including their development of an understanding of ethical principles. His theory is the basis for the Piaget-Kohlberg theory of ethical development.

Postconventional ethics The third and highest stage of ethical development in Dewey's theory of ethics. In this stage, individuals make decisions based on their own understanding of what is acceptable, exclusive of either personal gain or cultural mores.

Preconventional ethics The first and lowest stage of ethical development in Dewey's theory of ethics. In this stage, individuals make decisions based on reward and punishment as a result of actions. It is effectively a matter of seeking pleasure and avoiding pain.

Predictability A characteristic of existing in the mind state in which an individual's behavior lacks spontaneity, originality, or creativity and thus becomes highly predictable and determinable.

Prime directive A modern representation of the concept of reciprocity. Basically, it is the Golden Rule, that is, to love thy neighbor as thyself. In terms of ethics, it is the ultimate statement of what works.

Principled morality The higher stage of ethical decision making in the Piaget-Kohlberg theory of ethical development. In this stage, the individual finally takes personal responsibility for ethical decisions exclusive of what others may think and do. This is decision making based on the individual's understanding of ethical principles and right and wrong.

Punishment-obedience orientation The lowest stage of ethical decision making in the Piaget-Kohlberg theory of ethical development. In this stage, the individual determines behavior on the basis of obedience to rules and avoiding punishment. It is equivalent to Dewey's preconventional ethics.

Rational technology An approach to the development and implementation of technology that takes into consideration the effect that technology will have on all levels of humanity and on the world. It is rational because it realizes the interconnection among all the elements of the ecosystem.

Reciprocity One of the three universal laws of systemic behavior. According to the law of reciprocity, what you put into a system will determine what you get out of the system. In terms of ethics, what goes around also comes around. It is the central theme of determining what works.

Resentment The condition existing when we are reminded or faced with life issues as yet unresolved, particularly if there is a refusal to recognize the existence of such issues.

Responsible (response-able) Able to respond. Responsibility does not indicate obligation or limitation but rather an awareness of consequences that enhances the ability of the individual to act appropriately and in a way that works regardless of circumstances.

Revenge An act of violence perpetrated by one individual on another as a result of resentment and fear. Due to the reciprocal nature of reality, revenge always turns out to be against one's self.

Safety needs The second level of need fulfillment in Maslow's hierarchy of needs. In this level, an individual seeks to acquire safety and thus alleviate fear. It is only implemented, according to Maslow, after physiological needs are satisfied.

Self-actualization The highest level of need fulfillment in Maslow's hierarchy of needs. In this level, the individual achieves self-expression in the truest sense. It is an expression of integrity in which the highest understanding of the individual's being is expressed without fear. It is the point at which a person becomes genuine.

Self-esteem The fourth level of need fulfillment in Maslow's hierarchy of needs. In this level, the individual achieves a feeling of personal worth. True self-esteem is exclusive of the opinions of others and represents a stage of awe at one's own abilities, experienced without ego or fear.

Sermon on the Mount The primary expression of the teachings of Jesus and foundational tenets of the Christian faith. For purposes of an understanding of ethics, it is reflective of ethical behavior.

Shea, Gordon F. A twentieth-century ethicist who posits a level of ethical development called transcendent ethics, in which the individual realizes the connected nature of all life and the true nature of Karma.

Situational perfection The concept that at every point in time, the conditions that exist are perfect for what has gone before and represent a perfect picture of past behavior and actions. Recognizing situational perfection allows the individual to learn from experience and accept the truth of what is without judgment.

Social contract A concept in the Piaget-Kohlberg theory of ethical development. The social contract is an implied agreement among the members of society that they will behave in a way that is consistent with creating stability and a reasonable degree of satisfaction for everyone involved in that society.

Social needs The third level of need fulfillment in Maslow's hierarchy of needs. Such needs (belonging and those needs related to the social aspects of self-esteem) are directly dependent on group contact and membership.

Survival traits Species-specific traits that allow members of that species to survive in their environment.

Synergy The first of the universal laws of systemics. According to the law of synergy, the whole is greater than the sum of its parts. That is, there is an advantage to cooperation within a system that yields more than just the sum total of individual elements.

System A collection of elements that interact to achieve a goal or reach a given state.

Systemic structure The structure of all systems. Systemic structure consists of cooperating elements called subsystems made up of other elements that are also subsystems. By studying such a structure, it is theoretically possible to define the relationships among the various elements and understand the mechanisms by which the system operates.

***Tao Te Ching* (Teaching of Tao)** The book of Taoism. It is a collection of instructions describing The Way and how those on the road to perfection conduct themselves. It is attributed to Lao-tzu, though it is probable that a number of contributors were involved.

Taoism A religious philosophy of Asia that explains a mode of conduct that leads to perfection. This method of living, also known as The Way, is seen as the only valid purpose for a rational person, and though perfection is never actually reached, merely seeking it will continually increase the well-being of the individual.

Technologization The creation of technology by humans as a method of external evolution. It is an innate survival trait in humans.

Technology The collection of objects and methodology developed by humans to ensure their survival and success as a species.

Ten Commandments The primary expression of acceptable behavior of Judaism. It is a set of ten statements or laws describing what is and is not acceptable behavior in the sight of God. In terms of ethics, it represents ten statements of reciprocity.

Thoreau, Henry David A nineteenth-century transcendentalist who operated ethically on the basis of ethical principle rather than lower level understandings of ethical conduct. He was known as a nonviolent protestor firmly grounded in concepts of principled ethics.

Transcendent ethics An ethical state proposed by Gordon Shea in which an individual naturally does the ethical thing at all times by virtue of realizing the connected nature of all reality. In such a state, it is recognized that any action against another is an action against oneself because all are part of the whole, and therefore ethical conduct eliminates any possibility of harm to another.

Universal ethical principle orientation The highest level of ethical development in the Piaget-Kohlberg theory of ethical development. In this level, a person goes beyond principled ethics to realize that no act is ethical unless it is a positive act for all concerned. Thus there is a universal component to ethical acts, and if anyone is injured by an act, the act is not ethical in nature.

The Way The Tao. A process of discovering to a higher and higher degree the perfection of the universe in which life is a journey toward that understanding of perfection.

Wealth A state of being experienced by those not living in the mind state. In this state, everything wished for is available when needed. It is a realization of the abundance of the universe.

Wicca A traditional Celtic religion still in existence in many places in the world. Wicca centers on the principles of unity and reciprocity.

Witch A female practitioner of the Wiccan religion who uses spells, incantations, rituals, and natural herbs to affect change for others and herself.

Worldview Another name for paradigm. It is the manner in which an individual sees reality, which is individual and self-imposed through beliefs, attitudes, training and experiences, and cultural influences.

INDEX